A PRACTICAL GUIDE TO INSPECTING

HEATING & COOLING

By Roy Newcomer

CONTENTS

INTRODUCTION

My background includes many years in construction and several more as the owner of a Century 21 real estate franchise. In 1989, I started a home inspection company thatgrew larger than I ever imagined it could. Training my own staff of inspectors to the highest inspection standards led to my teaching home inspection seminars across the country and developing study courses, books, and videos for home inspectors. The American Home Inspectors Training Institute was founded as a result of my desire to share this experience and knowledge in home inspection.

The *Practical Guide to Inspecting* series is intended for both beginning and experienced home inspectors. So if you're studying home inspection for the first time or are using the materials as a refresher, these guides should be of assistance to you.

I've created these guides to include all aspects of home inspection. Not only a broad technical background in home systems, but the other things you need to know in order to perform a *good* inspection of those systems. They lay out technical information, guidelines for the inspection, how-to instructions for inspecting system components, and the defects, deficiencies, and problems you'll be looking for during the inspection. I've also included some advice on how to report your findings to the home inspection customer.

I've been a member of several professional organizations for a number of years, including ASHI® (American Society of Home Inspectors), NAHI™ (National Association of Home Inspectors), and CREIA® (California Real Estate Inspection Association). I am a great supporter of those organizations' quest to promote excellence in home inspection.

I do encourage you to follow the standards of the organization to which you might belong, or any state regulation that might take precedent over the standards used here. Use the standards in this book as a general guide for study and apply the standard or state regulation that applies to you.

The inspection guidelines presented in the Practical Guides are an attempt to meet or exceed standards and regulations as they exist at the revision date of the guides.

There's a lot to learn about home inspection. For beginning inspectors, there are some *hands-on exercises* in this guide that should be done. I'm a great believer in learning by doing, and I hope you'll try them. There are also some of my *personal inspection stories* to let you know what it's really like out there.

The *inspection photos*, referenced in the text can also be found at www.ahit.com/photos. You'll read the story about each one as you go along. Be sure to watch for my *Don't Ever Miss* lists. I've included them to alert home inspectors to report those defects (if found during the inspection) in the inspection report. If missed, these items are often the cause for lawsuits later. Finally, to help you see how you're doing as you study this guide, I've included some *worksheets.* The answers are given for each one for self checking. Give them a try. Checking yourself can help you lock important information in your mind. There's also a *final exam* that you can complete and send in to us. Many organizations and states have approved this book for continuing education credits. Submit the exam with the required fee if you need these credits.

In total, the *Practical Guide to Inspection* series covers all aspects of the general home inspection. Each guide covers a major aspect of the inspection, as their titles show:

Electrical
Exteriors
Heating and Cooling
Interiors, Insulation, Ventilation
Plumbing
Roofs
Structure

If you are interested in other titles in the series, please call us at the American Home Inspectors Training Institute to order them. Call toll free at 1-800-441-9411.

Roy Newcomer

INSPECTING
HEATING
AND
COOLING

Chapter One

THE HEATING INSPECTION

Chapter One presents the inspection of the home's heating system. In general, there are two kinds of home heating — central heating and area heating. The home inspection concerns itself with *permanently* installed heating systems, both central and area. The basic components of central heating systems include the following:

- A **safe compartment** in which to convert fuel or energy to heat. This includes a burner and combustion chamber for converting fossil fuels such as gas or oil to heat or a chamber containing a resistance coil for converting electrical energy to heat. When gas or oil is used, air must be supplied to the chamber for combustion purposes.

- A **heat exchanger** for transferring this heat to the air or to water. The heat exchanger that transfers heat to air is called a **furnace**; one that transfers heat to water is called a **boiler**.

- For fossil fuel furnaces and boilers, a **disposal system** of flues, vents, and chimneys to remove combustion products from the home.

- A **distribution system** of ducts or pipes that carry warm air, hot water, or steam throughout the house. Warm air may be conveyed through the distribution system naturally by gravity or pushed by a circulating fan. Hot water can be conveyed by gravity or set in motion by a circulating pump.

- **Heat outlets** such as registers or radiators for transferring heat into each room.

- **Temperature and safety controls**. Each central heating system requires a thermostat that turns the furnace or boiler on and off in order to supply heat as needed. And each is designed with automatic (and manual) safety controls that will turn off heating equipment when the equipment malfunctions.

A home may be heated entirely with electric baseboard or wall-mounted area heaters or radiant panels. In this case, heat is generated within each room to be heated, not centrally through a distribution system.

Guide Note

Pages 1 to 8 outline the content and scope of the heating inspection. It's an overview of the inspection, including what to observe, what to describe, and what specific actions to take during the inspection. Study these guidelines carefully. These pages also present some special cautions about inspecting the heating system. Please read them and be aware of the potential of harming yourself or damaging the equipment.

For Your Library

In this guide, we're going to be suggesting a few books on heating systems that will be of benefit to you as a home inspector. Start now and build a good library of reference books. The following book is a good general overview of heating and air conditioning systems.

Home Heating and Air Conditioning Systems by James L. Kittle. Published by TAB Books in 1990. Available in most bookstores for $16.95.

THE HEATING INSPECTION

- Heating equipment, operating and safety controls
- Combustion product disposal system
- Distribution system
- Heat source per room

Inspection Guidelines and Overview

These are the guidelines that govern the inspection of the heating system. Please study them carefully.

Heating System		
OBJECTIVE		To identify major deficiencies in the central heating system which do not require detailed heat-loss analyses.
OBSERVATION		<u>Required to inspect and report:</u> • Permanently installed heating systems — Heating equipment — Normal operating controls — Automatic safety controls — Chimneys, flues, and vents — Solid fuel heating systems • Heat distribution systems — Pipes, fans, ducts and piping, with supports, dampers, insulation, air filters, registers, radiators, fan-coil units, convectors • The presence of an installed heat source in each room • Energy source • Main fuel shutoff <u>Not required to observe:</u> • Interior of flues • Fireplace insert flue connections • Humidifiers • Electronic air filters • The uniformity or adequacy of heat supply to various rooms
ACTION		<u>Required to:</u> • Operate the system using normal operating controls. • Open readily openable access panels provided by the manufacturer. <u>Not required to:</u> • Operate heating systems when conditions may cause equipment damage. • Operate automatic safety controls. • Ignite or extinguish solid fuel fires.

Not every detail of what is to be inspected and what is to be reported is stated in these standards. Consider this chart a solid outline of what is required. There are many, many other details you'll learn in this Study Unit. Here's an overview of the heating inspection:

- **Heating equipment:** The home inspector inspects the main heating unit in the home — furnace, boiler, or area units — recording the **brand name** of the unit, the **type** of heating unit such as forced warm air or gravity hot water, and its **fuel source** such as gas, oil, or electricity. The inspector is required to **open readily openable manufacturer's access panels** during the visual inspection and to **operate the heating system using normal operating controls** to check the operation of the heating unit. However, the inspector is not required to operate heating systems that are shut down, that don't respond to normal operating controls, or when conditions may cause equipment damage.

The condition of the unit's fuel source equipment, outer jacket, the burner and combustion chamber, and the visible portions of the heat exchanger are all inspected. Cracked heat exchangers are reported as a **safety hazard** and as a **major repair** since furnace or boiler replacement is recommended. The inspector is **required to observe, but not operate, automatic safety controls**.

The home inspector also estimates the unit's age and its **remaining useful lifetime**. It's important to know whether a furnace or boiler can be expected to need replacement within the next 5 years and to record that fact in your inspection report. The following chart gives an overview of the life expectancy of various furnaces and boilers.

Heating System	Life Expectancy
Gas-fired warm air furnace	15 to 25 years
Oil-fired warm air furnace	20 to 30 years
Electric systems	20 to 25 years
Cast iron boiler	30 to 50 years or more
Steel boiler	30 to 40 years or more
Copper boiler	10 to 20 years
Circulating pump (hot water)	10 to 15 years

For Your Library

The Preston's Guide to HVAC lists most brand names and models of furnaces and air conditioning systems, including the date of manufacture. It's an indispensable tool for determining year of manufacture and estimating the remaining lifetime of a heating system, and every home inspector should have one.

Guide Note

The inspection of fireplaces and wood stoves are presented in another of our guides — A Practical Guide to Inspecting Interiors, Insulation, Ventilation.

NOTE: The home inspector is required to inspect **solid fuel heating systems**, referring to fireplaces and wood stoves. These items will not be covered in this guide. Fireplaces and wood stoves are presented in *A Practical Guide to Inspecting Interiors, Insulation, Ventilation.* As stated in the standards, the home inspector is **not required to ignite or extinguish solid fuel fires or to observe fireplace insert flue connections**.

- **Combustion product disposal system:** For oil and gas-fueled furnaces and boilers, the home inspector inspects chimneys, flues (smoke pipes), and vents for proper installation and safety. Evidence of combustion products leaking into the home is reported as a **safety hazard**. Draft diverters are examined, and the smoke pipe is checked for the operation of dampers, corrosion, pitch, proper supports, and the seal at the chimney. The chimney cleanout is investigated to determine whether the chimney is blocked. The home inspector **is not required to inspect the interior of flues**.

- **Heating distribution system:** The inspection of the duct or piping system includes observing the operation and condition of fans and circulating pumps. Air filters are inspected for warm air systems. Although most standards state the home inspector is **not required to inspect electronic air filters**, we suggest that you do — carefully, with the proper cautions in mind. The home inspector checks humidifiers as far as their general condition and effect on other heating equipment, but is **not required to inspect the operation of the humidifier**.

  The home inspector checks all visible **ducts and piping** for proper support, dampers, leaking (in hot water and steam systems), and insulation. The home inspector checks the condition and operation of all **heat outlets** such as registers, radiators, fan-coil units, and convectors.

- **Heat source per room:** A seemingly minor point, but important to customers, is the presence of a heat source in each living area. The home inspector checks for a heat source in each room and records its presence or absence in the inspection report. However, the home inspector is **not required to determine or report on the adequacy of the heat supply to each room**. The home inspector also pays attention to the location of cold air returns with warm air systems.

Inspection Equipment

There are some basic inspection tools that are necessary for the inspection of the heating system. The home inspector needs an **inspection mirror** when examining the heat exchanger from all possible angles. The

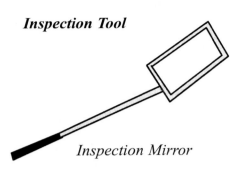

Inspection Tool

Inspection Mirror

inspection mirror has a long handle and a relatively small mirror face. Also important is a **high-powered flashlight** to light up the dark areas while you use the mirror. A **6-in-1 screwdriver** will come in handy for removing furnace access panels. However, some panels are bolted on, so it would be a good idea to purchase a set of **small wrenches** to keep with your inspection equipment. A **digital thermometer**, will come in handy when performing a temperature rise test in heating systems and a temperature differential test in cooling systems.

Some inspectors purchase a special **combustible gas detector** to check for combustion byproducts such as carbon monoxide around the heat exchanger and draft hoods, diverters, and dampers on the flue outlet. In this guide, we will present techniques for inspecting the heat exchanger that involve flame observation and visual checking with the inspection mirror and flashlight. And, for checking around the flue outlet, we'll describe a test using simple **matches**. Although these tests are acceptable for home inspection, our own home inspectors also use a combustible gas detector, and we recommend that you add it to your list of tools.

Inspection Concerns

In some cases, inspecting the heating equipment can be dangerous. There are 3 concerns that the home inspector should keep in mind during the inspection of the heating equipment — the customer's safety, the inspector's own safety, and damage to the heating equipment. Here are some general rules about the heating inspection that the home inspector should follow:

- **Keep your customer safe.** Of greatest concern during any part of the home inspection is the customer's safety. The home inspector is responsible for the well-being of the

INSPECTION TOOLS
• Inspection mirror
• Screwdrivers
• Wrenches
• High-powered flashlight
• Thermometer
• Matches
• TIF8800 combustible gas detector, if desired

"An instructor here at the American Home Inspectors Training Institute tells how he and a customer were kneeling in front of a gas warm air furnace waiting for it to fire. When the pilot didn't light right away, he knew they were in trouble.

"He pushed his customer to the floor away from the furnace and dove to the floor himself. When the pilot ignited, a flame shot out 4 1/2' from the furnace. It would have burned them both.

"Now he most vigorously cautions all of our students not to do what he did. Always keep yourself and your customer out of the way when firing the furnace."

Roy Newcomer

customer. The #1 rule of home inspection is to have the customer present during the home inspection, and that does mean during the heating inspection too.

However, there is a moment of danger during the heating inspection when you want to pay close attention to where your customer is. That's **when the heating unit is first fired up**, especially with gas-fired furnaces and boilers. With a gas-fired system, there can be too long of a delay before the burner fires while gas is pouring into the combustion chamber. Then when the burner fires, the condition causes a huge flameout, shooting a flame a few feet out from the furnace. When firing the unit, both you and your customer should stand to the side. Don't ever let the customer peer into the chamber at this time. And don't you either.

- **Protect yourself from harm.** Some of the standards of practice specifically state observations and actions that the inspector is *not* required to perform. These standards were developed in the interest of inspector safety as well as to prevent the inspector from damaging the equipment. The standards state that the home inspector is not required to inspect the interior of flues, humidifiers, and electronic air filters.

- **Do not disassemble flue vent piping** during the heating inspection. There is a risk from harmful filth that can be present in the flue and a possibility that the piping can't be reassembled. There are other ways to detect a blocked flue other than taking it apart. You can look through the barometric damper or check the chimney cleanout with your inspection mirror.

The home inspector will be checking for leaking and corroded humidifiers and their effect on the condition of the furnace and ducts, but is not required to check their operation. **Don't disconnect humidifiers** to obtain access to furnace plenums. Too many inspectors have been cut by sheet metal doing this.

Although some standards say not to inspect **electronic air filters**, we're going to suggest that you do — with caution. The electronic or electrostatic air filter contains a series of fine wire grids that are given an electrostatic charge, which aids the filter in capturing small particles of dust, smoke, and pollen. When the filter is working, it pops and crackles like a bug zapper. The danger to the inspector when testing this type of filter is **electrical shock**. To prevent shock, turn off the master furnace switch which will turn off power to the filter, then wait about 30 seconds to let the static charge dissipate before pulling the filter out for inspection.

When inspecting boilers, **don't operate the pressure relief valve** and be sure not to catch your sleeve on one to trip it. Hot water or steam spraying into your face is something you don't want to experience.

- **Don't cause damage to the equipment.** The home inspector's job is to point out defects in the heating system, not to cause them. **Don't dismantle equipment** other than what you'll be instructed to do in this guide — no matter how curious you may be or if the customer requests you to. Too many things can go wrong. When you reassemble equipment, it may not work again.

You'll be inspecting heating units for rust and corrosion, both inside and outside. **Don't ever scrape at or pick at corrosion** with your screwdriver. This can cause the rust or corrosion to fall away, opening holes in pipes, ducts, burners, and heat exchangers. Simply report the condition. Don't make holes.

Be careful too with **heating ducts** when inspecting for proper supports. You don't need to hang on them and pull them down to find out if they're not properly supported.

- **Under certain conditions, don't operate the heating equipment.** Some conditions can cause equipment damage or an unsafe situation in the home. Do *not* turn on equipment that:

  — **Has been shut down.** The system may be shut down for a good reason such as open piping, unsafe wiring, leaks, an unsafe chimney, and fire risks.

  — **Has been switched off.** When a heating system is turned off, always talk to the owner or a responsible party and get their permission to turn it on. It may be off because of one of the reasons stated above, or you may get permission to go ahead. That's okay. But always check first.

  — **Does not appear to be vented properly.** If you notice this condition before firing, don't turn the equipment on. If you've already fired the unit and you observe

---

IMPORTANT RULES

- Keep your customer safe.
- Protect yourself from harm.
- Don't cause damage to the equipment.
- Don't operate equipment that has been shut down, switched off, is improperly vented, has a suspect chimney, or indicates an unusually high pressure or temperature.

Photo #1 shows **handyman work** *on a furnace. We came across this furnace wrapped in aluminum foil. This was the owner's idea of trying to conserve heat. Study the photo to see what else you can see. The owner didn't want so much heat going up the chimney, so he diverted hot combustion gases from the smoke pipe back to the furnace cold air return (piping coming back from the flue at the right). The owner was pumping exhaust back through the house. Now that's improper venting in a big way and terribly dangerous. We didn't fire up this furnace.*

Personal Note

"Here at our offices, we call the house in Photo #1 the <u>dead-guy house</u>.
"One of my inspectors did this inspection for the estate of an old man who had recently died. After studying the furnace and realizing the old man had re-routed exhaust gases back into the furnace, the inspector asked how the man had died. The answer was they didn't really know —old age they thought, his heart just stopped. 'Try carbon monoxide poisoning,' our inspector said.
"He did not fire up that furnace, and he gave a severe warning, reporting the handyman venting as a deadly safety hazard. Of course, nothing was done at that late date to try to determine the old man's cause of death, but we're convinced it was CO poisoning."

Roy Newcomer

improper venting or an apparently blocked flue, turn the equipment off promptly.

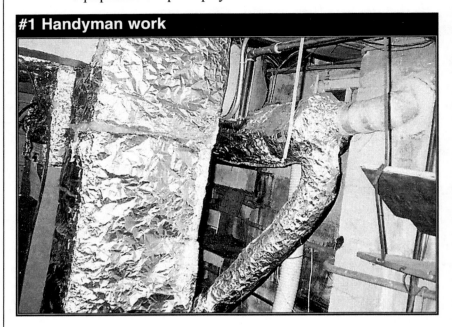

#1 Handyman work

— **Has a suspect chimney.** For example, you may find an old single wythe, unlined brick chimney with visible damage outside and in the attic. It wouldn't be safe to fire up the heating unit in this case.

— **When gauges show unusual pressure or temperature conditions.** If after firing up a boiler, you notice pressure and temperature conditions beyond a safe level, do not let the boiler continue to operate. Turn off the equipment immediately.

— **A steam boiler's water gauge shows no water.** If the water level is too low in a steam boiler, operating the system can cause the boiler to overheat and crack. Do not operate the system.

Chapter Two

GENERAL INFORMATION

This chapter will present some general heating terms and concepts the home inspector should be familiar with.

Thermostats

The main operating control for a heating system is the thermostat, a temperature-sensitive device that opens and closes a circuit in response to changes in temperature. A thermostat normally operates at low voltage (24 volts), although some operate at line voltage (generally used with electric resistance heating). The older thermostats simply start the heating system on a call for heat and turn it off when the thermostat is satisfied.  When the temperature near the thermostat falls below a preset temperature, the contacts within the thermostat close and activate the heating system. When the nearby temperature rises to the preset temperature, the contacts open again and shut it down.

In some thermostats, the contacts are exposed to the air. A layer of dust can accumulate on the contacts and interfere with normal operation. They can also burn or corrode. In modern thermostats, the contacts are enclosed in glass or have a sealed mercury switch in which a drop of mercury makes and breaks the circuit as it swings back and forth.

There are a variety of thermostats available today. Many newer models have **setback features**. Electronic thermostats can be set to lower the temperature setting automatically at night and raise it again each morning. Generally, the temperature differential with this setback features is from 5° to 10°. Some can be programmed for double setback, where the homeowner can also have the heat raised and lowered automatically during the daytime as well as nighttime hours.

Guide Note

Pages 9 to 17 present an introduction to general heating terms, concepts, and types of heating systems.

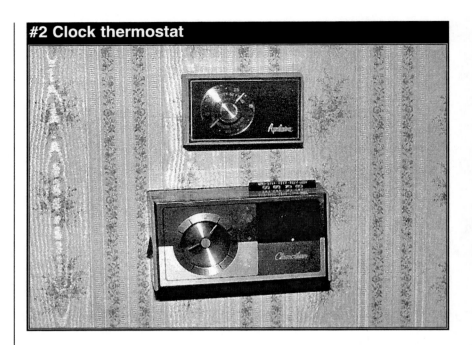

#2 Clock thermostat

*Photo #2 shows a **clock thermostat** (below) with a nighttime setback feature.*

#3 Electronically programmable model

*Photo #3 shows an **electronically programmable model** with setback features. It should be noted that this thermostat has batteries in it. If the safety switch to the heating system is turned off and the batteries are low in the thermostat, it's possible to lose its programming.*

Most thermostats have a **fan control** for forced-air systems. This control can be set so the blower operates automatically when the furnace is on or operates continuously for constant air circulation.

Most thermostats have an **anticipator**, which anticipates the preset point on the thermostat and turns the heating system on before the preset temperature is reached. This compensates for the lag time between closing the circuit and delivering the needed heat. The anticipator will also turn off the heating system just before the desired temperature is reached, letting the remaining heat in the combustion chamber bring the temperature up those last few degrees. The anticipator prevents **overshooting**, which occurs when remaining heat in the combustion chamber heats the house higher than the preset temperature after the burner is turned off.

The anticipator should be calibrated to the particular heating system in the house. If the anticipator is not set correctly, a condition called **short cycling** can occur. That's when the anticipator shuts down the system too soon, so the desired temperature is never reached. Then the heating unit will immediately kick on again, only to shut down too soon again. (There can be other causes of short cycling.)

During the inspection of the heating system, the home inspector should **operate each thermostat** to be sure it turns the heating system on and off. But be sure to note the temperature setting before you test the thermostat, so you can return it to the same setting again. When inspecting thermostats, watch for the following conditions:

- **Improper location:** Thermostats should be located on inside walls of the home. They should be in locations where factors other than normal home conditions won't unduly affect the temperature in the area. For example, thermostats in cold drafts or too near the front door can turn the heating system on too often. Those located in the path of direct sunshine or fireplace flames won't turn the heating system on often enough.

- **Loose, unlevel:** If the thermostat is loose on the wall or unlevel, the mercury can open and close the circuit inaccurately, and the calibrations won't correspond to the actual temperature. Some home inspectors carry a thermometer and test the actual temperature against the thermostat calibrations. If differences are found, they should be reported. Customers should be advised that loose or unlevel thermostats should be fixed.

- **Not working:** If the heating system doesn't kick in when the thermostat is operated, it can be for several reasons. The system may be turned off. Obviously, if the system doesn't come on, the situation should be further investigated to determine if the problem is with the system

Personal Note

"I once had a problem with a programmable thermostat. Its batteries were low, which I had no way of telling. When I turned off the safety switch to the furnace, the program in the thermostat was lost. The homeowner soon discovered the problem and had to reprogram the thermostat.
"When I see this type of thermostat, I tell the homeowners about the low-battery situation and suggest that they check whether the program is still functioning after the inspection. That's better than getting a complaint call later."
Roy Newcomer

Personal Note

"During a home inspection I performed, I turned up the thermostat but the furnace wouldn't fire. I couldn't figure out why. We called in a heating contractor to take a look at the heating system. He took off the thermostat cover, scraped nicotine off the contacts, and the furnace fired.
"Thermostats with unprotected contacts can cause problems, especially if there are smokers in the house."

Roy Newcomer

Definitions

The <u>master shut-off</u>, also called the <u>safety switch</u>, turns off electricity to the heating system and its controls. The <u>serviceman's switch</u> serves the same purpose as the safety switch and may be the same switch.

or the thermostat. It can happen that dirt on the open contacts in the thermostat prohibits the circuit from closing. Open contacts should be cleaned regularly. Thermostats can suffer mechanical damage or simply fail. These should be replaced.

Zoning

Some homes have zoned heating, where each section or zone of the home is controlled by a separate thermostat. The term **zone control** means different areas of the house are under the control of different thermostats. Sometimes, zone control is achieved by installing totally independent furnaces or boilers for upstairs and downstairs heating, each with its own thermostat. Electric baseboard heating and wall mounted strips may be installed as separate units in each room, controlled by wall thermostats or by elements mounted on the baseboard case itself.

A single heating system can be designed for zoned heating. With hot water systems, the boiler can have different circulators or zone control valves to provide heat to different parts of the home. With forced warm air systems, heat distribution can be controlled by motorized dampers in the heat ducts for various areas of the home.

Master Shut-Off

Another normal operating control for central heating systems is a master shut-off or **safety switch** with which to turn off the heating system by hand in the case of an emergency. Safety switches turn off electricity to the furnace or boiler and its controls. When the switch is off, adjusting the thermostat will no longer allow the equipment to fire.

With oil-fired heating equipment, a **remote safety switch** is required, usually at the top of the basement stairs or outside the furnace or boiler room, so the burner can be turned off without having to approach it. Remote switches often have a red cover plate to distinguish them from others. Gas-fired systems have safety switches either near the equipment or at a remote location, as is required in some areas of the country. With electric furnaces or boilers, its circuit breaker in a nearby electrical panel may serve as the safety switch.

The home inspector should **test the safety switch** during the inspection of the furnace or boiler. When this master safety switch is turned off, the heating burner should stop. Any defective switches should be reported. Some local areas require the boiler's circulating pump or the furnace's fan to remain in operation. The home inspector should find out what his or her local requirements are.

CAUTION: Always be sure that the safety switch is turned back on after an inspection. There are cases of inspectors leaving an empty house without heat during the winter and the condition not being discovered until too late — after plumbing pipes have frozen and burst.

There is also a **serviceman's switch** within easy reach of the heating unit which serves the same purpose as the safety switch. With gas and electric equipment, the safety and serviceman switch may be one and the same, although not with oil-fired furnaces and boilers.

Clearances and Codes

When inspecting the heating system, the home inspector should pay attention to the clearances around the heating unit itself and around exhaust pipes for oil and gas equipment.

Various models of furnaces and boilers may have different **clearance from combustibles** requirements, but typical requirements are at least a 6" clearance above the unit and 0-6" on all sides except the front of the unit. For oil-fired furnaces and boilers, a clearance of 24" is required between front of the unit and nearby combustibles, while gas and electric units may require as little as 0". The exhaust pipe requirements vary too, depending on type of fuel and materials used. A single-wall metal pipe requires clearance of 18" for either oil or gas furnaces or boilers <u>without a draft hood</u>. With a draft hood, a 9" clearance is required for oil, and a 6" clearance for gas. Double-wall pipes need clearances of only 2", while a 1" clearance is enough for Type B gas vents.

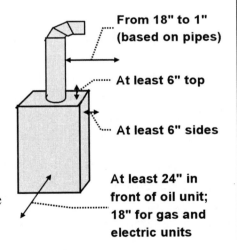

From 18" to 1" (based on pipes)

At least 6" top

At least 6" sides

At least 24" in front of oil unit; 18" for gas and electric units

---

A CAUTION

Always be sure that the <u>safety switch</u> for heating equipment is turned back on after you test it. Check before you leave that the heat is back on.

CLEARANCE FROM COMBUSTIBLES

- 6" on top and sides of furnace or boiler
- 24" front for oil-fired furnace or boiler
- 18" front for gas and electric units
- Single-wall smoke pipe: 18" for units without draft hoods, 9" for oil with draft hood, 6" for gas
- 2" for double-wall smoke pipe
- 1" for Type B gas flue pipe

A BTU is a measure of heat output. BTU stands for British Thermal Unit and 1 BTU represents the amount of heat required to raise the temperature of 1 pound of water 1° Fahrenheit.

Exhaust or **smoke pipes** should slope upwards towards the chimney at a slope of 1/4" per foot of length and be well supported. Smoke pipes should be as short as possible. Local codes give requirements about length and the number of elbows (bends) allowed in the run.

Chimney clearance typically follows the 3-2-10 rule. That is, the chimney should extend 3' above the roof's surface, and it should be 2' higher than anything within 10' of it, including the ridge, dormers, parapets, and so on. If a new

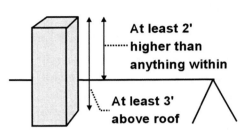

At least 2' higher than anything within

At least 3' above roof

addition is built that is higher than the existing chimney, the chimney must be extended to meet these requirements. Chimneys not constructed to these standards can experience back draft problems and spillage.

Fossil fuel heating units (oil or gas-fired, for example) must rely on a supply of combustion and draft air for safety. When units are located in **closets or enclosed rooms**, codes require 1 square inch of ventilation per 1000 BTU's of output. (A **BTU** is a British Thermal Unit, and 1 BTU represents the amount of heat required to raise the temperature of 1 pound of water 1° Fahrenheit.) Closet doors should be louvered,and/or rooms should be vented at the top and bottom.

Any fuel burning furnace or boiler located in the **garage**, should sit with the burner at least 18" above the floor. This allows gasoline vapors, which hug the floor, to dilute before reaching the burner level.

HEAT TRANSFER

- Conduction by physical contact
- Convection by warming air, water, or steam and moving it to a cooler location
- Radiation by radiating energy

Heat Transfer

The concept of heat transfer refers to transferring heat from a warmer object to a cooler object. This happens in 3 basic ways:

- **By conduction:** With conduction, heat is transferred from a warmer object to a cooler object by physical contact. An example of conduction is when you touch a hot stove and burn your finger. Home heating is not accomplished by conduction.

- **By convection:** Most homes are heated with convection

heating. With convection heat transfer, heat is transferred by a fluid such as air, water, or steam that can absorb the heat from a warm location and move it to a cooler location. With forced air systems, warm air arrives in the living space through **registers**, warming the air in rooms and consequently the people in those rooms.

When hot water heating systems send hot water to **convectors**, the air in contact with the convectors becomes warmed, and as a result so do the people in the warm air. A convector is a terminal unit or heat outlet, commonly a baseboard unit, consisting of flat plates, finned plates, or finned pipes over which air passes and is warmed. This is also convection heating.

- **By radiation:** With radiation, heat is transferred from a warmer object to a cooler object (or body) that is *not* in physical contact with it. Radiation heat does not warm the air itself. It turns to heat only when the energy is absorbed by the cooler body. An example of radiation heat transfer is when you sit next to a hot stove and the heat is transferred to you from the hot metal.

Hot water and steam heating systems move heated water or steam to metal **radiators** that radiate heat energy in all directions. People in the vicinity absorb the energy, transforming that energy into heat, and are thereby warmed.

To a certain extent, radiators also heat by conduction, in that the air in contact with the radiator will also be heated.

There are **radiant heating systems** that, in effect, use the entire floor, wall, or ceiling as a radiator. Buried pipes or electric cables heat the surface by conduction; people in these rooms are then heated by the heat energy radiated from the surface.

Types of Heating Systems

This guide will present information on various types of heating systems the home inspector is required to inspect in the course of the heating inspection. The basic types of heating systems are as follows:

- **Gravity warm air:** The gravity warm air furnace heats air that rises naturally from the furnace through a duct system and circulates throughout the house. Cooler air falls

Definitions

Conduction is the transfer of heat from a warmer object to a cooler object by physical contact.

Convection is the transfer of heat through air, water, or steam that moves heat from a warmer location to a cooler location. Convection heating systems ultimately heat the air in the living space.

Radiation is the transfer of heat from a warmer object to a cooler object not in contact with it by radiating heat energy.

HEATING SYSTEMS

- Gravity warm air
- Forced warm air
- Gravity hot water
- Forced hot water
- Steam
- Electric resistance
- Radiant
- Heat pump

through a return duct or opening in the floor. No motorized blowers are used. Fuel can be coal, oil, or gas.

- **Forced warm air:** Heat produced by the fuel warms the heat exchanger. Air circulation is achieved when a blower, sensing the warm air available, blows the warm air into a duct system to registers throughout the house. Cold air return ducts bring air back to the furnace, completing the cycle. These furnaces are typically fueled by oil, gas, or electric coils. Solar energy may also be used.

Much progress has been made with forced warm air systems in recent years and **high efficiency furnaces** are available that perform to almost 100% efficiency. They are based on the principle of cooling combustion gases to below their dew point, condensing the vapor and recovering as much latent heat as possible. This requires that furnaces use outside air for combustion, have 2 heat exchangers to extract the extra heat, and have a drain to dispose of combustion gas condensate. They may also have an unconventional combustion system.

- **Gravity hot water:** This system relies on the principle that hot water will rise naturally, pushing the cooler water ahead of it. Once the boiler is fired, hot water starts to move through a pipe system to radiators in each room that give off heat. Cooler water returns to the boiler. Gravity hot water systems are fueled by coal, oil, or gas.

- **Forced hot water:** This system uses a circulating pump to push the hot water through the piping to radiators or convectors in each room. Cooler water returns to the boiler through the same pipe or a second piping system. The system is typically fueled by coal, oil, or gas. Newer high efficiency boilers are available, based on the same innovations used in high efficiency furnaces.

- **Steam:** Here, water is heated in the boiler to produce steam that rises naturally through a piping system to radiators in the house. When cooled, the steam condenses in the pipes and returns as water to the boiler. These units are heated by gas, oil, wood or pellets and earlier by coal.

- **Electric resistance heating:** Another method of heating is to have baseboard units or panels on walls or ceilings in each room. These resistance heaters contain a circuit to generate heat.

- **Radiant heating:** Radiant heating consists of electric cables, pipes or tubing buried in floors or ceilings during construction. Electric cables or hot water in the buried pipes heat the surface and radiate heat to the living space.

- **Heat pumps:** **Air-to-air** heat pumps are air conditioners capable of reversing the flow of refrigerant. Because a gas heats when compressed and cools as it expands, air conditioners work by compressing a refrigerant and expelling the resulting heat outdoors. Then the refrigerant expands and the resulting cold air is expelled into the house. With heat pumps, this cycle is reversed — expelling cold air outdoors and sending hot air into the house through ductwork.

WORKSHEET

Test yourself on the following questions.
Answers appear on page 20.

1. According to most standards of practice, the home inspector is required to:

   A. Operate automatic safety controls.
   B. Identify major deficiencies in the central heating system.
   C. Observe the uniformity or adequacy of heat supply to various rooms.
   D. Ignite or extinguish solid fuel fires.

2. When inspecting the heating equipment, the inspector is <u>not</u> required to:

   A. Operate the system using normal operating controls.
   B. Open readily openable access panels provided by the manufacturer.
   C. Observe automatic safety controls.
   D. Operate heating systems when conditions may cause equipment damage.

3. When should the home inspector have the customer stand back for safety's sake?

   A. When the thermostat is manipulated
   B. When the venting is inspected
   C. When the furnace or boiler is fired
   D. When the safety switch is tested

4. What should be a concern when inspecting electronic air filters?

   A. Electrical shock
   B. Harmful filth
   C. Cuts from sheet metal
   D. Explosion

5. If a heating system appears to be improperly vented after the home inspector fires it, the inspector should:

   A. Investigate the cause of improper venting.
   B. Turn off the system promptly.
   C. Ask the owner to inspect the venting.

6. How do setback features function on a thermostat?

   A. They anticipate the preset temperature and turn the heat on before that temperature is reached.
   B. They allow the owner to set the thermostat for a lower temperature setting for certain time periods.

7. What is overshooting?

   A. A condition where a house is heated higher than the thermostat is set for
   B. A condition where the heating system turns on and off too frequently
   C. A condition where different zones of a house are controlled by different thermostats
   D. A condition where the thermostat is unlevel causing miscalibration of the temperature readings

8. What type of central heating system is <u>always</u> required to have a remote safety switch?

   A. Gas-fired
   B. Oil-fired
   C. Electric

9. Forced warm air heating systems are an example of what type of heat transfer?

   A. Conduction
   B. Convection
   C. Radiation

10. What should the inspector be sure to do when the heating inspection is finished?

   A. Turn off the heating equipment.
   B. Set heating controls to the most efficient settings.
   C. Return controls to their original settings.

Chapter Three

GAS-FIRED SYSTEMS

In this chapter, we're going to discuss the principles, components, and the inspection of the gas components of gas-fired heating systems, regardless of the type of heating system the burners serve. The study of particular heating systems (warm air, hot water, steam, and so on) will begin on page 46.

A Word about Efficiencies

Before we discuss the components and inspection of gas-fired heating systems, let's sort out a few terms:

- **Conventional gas systems:** A conventional gas-fired furnace or boiler is the standard type available up to the mid-1970's before improvements began to be made. With these conventional heating systems, about 80% of the heat produced from burning gas stays in the house, while 20% goes up the chimney. Another 20% of the warm house air escapes up the chimney when the furnace or boiler isn't operating, and some additional fuel is wasted keeping the pilot burning. So, the conventional heating system is said to be between 55% and 65% efficient. With conventional systems, house air is used for combustion, the combustion chamber is unsealed, and warm exhaust gases rise naturally up the chimney.

- **Mid-efficiency gas systems:** Beginning in the mid 1970's, some modifications were made to conventional gas-fired furnaces and boilers that increased efficiency to the 80% range. One modification is the use of a **motorized vent damper** which closes when the burners were not in operation, preventing heat from escaping up the chimney. Another modification may be the use of the **induced draft fan** on the exhaust side of the furnace or boiler to pull combustion products through the unit during operation. The fan, when at rest, reduces heat loss when the burners are off. Most mid-efficiency systems use an **intermittent pilot** that lights by spark or hot surface ingitors, only when heat is called for. Basically, a mid-efficiency system is a conventional system with the modifications mentioned here.

Guide Note

Pages 19 to 33 present the study and inspection of the gas components of gas-fired heating systems.

	Efficiency
Conventional	55%-65%
Mid-efficiency	80%
High-efficiency	95%

- **High-efficiency gas systems:** The newest gas-fired furnaces and boilers can have an efficiency rating in the 95% range. The principle behind these systems, also called **condensing furnaces**, is to remove so much heat from exhaust gases that the gases are reduced to condensation. This is accomplished by having more than one heat exchanger. With high-efficiency units, the condensate is drained to the floor drain through plastic piping and cool gases exhausted to the outside through the house wall; no chimney is used. Other features of high-efficiency units are the use of outside air as combustion air, the presence of induced draft fans, and electronic ignition rather than a standing pilot.

In general, the home inspector has more access for the inspection of the components in conventional and mid-efficiency gas-fired heating units than with the high-efficiency units. We'll discuss these conventional components in greater detail.

Gas Supply

The gas that fuels the heating system may be **natural gas**, which is drawn from underground fields and purified and blended before being piped to the home by the gas company. **Bottled or LP** (liquefied propane or petroleum) **gas** can be used to fuel heating systems too. LP gas can be a single gas or a mixture such as propane, butane, or iso-butane.

Natural gas is distributed to the house in underground pipes at a high pressure of 90 psi (pounds per square inch) or an intermediate pressure of 30 psi. At the house side of the meter the gas pressure is reduced to **less than 0.25 psi** for home use. There may be a **pressure relief valve or regulator** before the meter to allow pressurized gas to escape. This regulator should be on the exterior of the house or vented to the exterior.

The home inspector can check the gas meter to see if it is adequately sized to supply gas to the heating unit and water heater. The meter is rated in cubic feet per hour. **One cubic foot of gas is approximately equal to 1,000 BTU,** so a meter rated at 250 cubic feet per hour will supply gas that is the equivalent of 250,000 BTU per hour. The BTU capacity at the meter should be equal to or greater than the sum of the input BTU requirements for the heating system and water heater.

Worksheet Answers (page 18)

1. B
2. D
3. C
4. A
5. B
6. B
7. A
8. B
9. B
10. C

Black iron gas piping of 1" or 3/4" is typically used to carry gas from the meter to the heating unit. It runs on the underside of the floor joists and is attached to the joists in a basement installation. The piping will drop down beside the heating unit to the level of the burners. Copper piping may be permitted for bottled gas lines in some communities. In some areas, flexible copper piping may be permitted for natural gas connections between metal lines and appliances. Plastic may or not be allowed. Cast iron or rubber lines are not allowed. Corrugated stainless steel tubing (CSST) a relatively new product brings gas from a manifold near the meter to the various gas-fired appliances throughout the house.

There should be a **manual turn-off valve** in the line at the heating unit so that gas supply can be turned off, if necessary, for maintenance work or an emergency. Where the gas line turns to meet the heating unit, a **drip leg** (also called a pipe trap) is installed to trap sediment and metal chips from the pipeline. (drip legs may be required depending on the moisture content of the gas). The gas line feeds into a **combination control** (automatic gas valve and pilot control valve) that is part of the heating unit. Gas is delivered to the burner through the gas **manifold**.

```
┌─────────────────────────┐
│        GAS LINE         │
│                         │
│  • Incoming black iron  │
│    piping               │
│  • Manual turn-off      │
│  • Drip leg             │
│  • Main gas valves      │
│  • Manifold to burner   │
└─────────────────────────┘
```

For Your Information

Check your local codes for what's allowed for gas lines in the home. See if copper, flexible copper, CSST, and plastic are permitted.

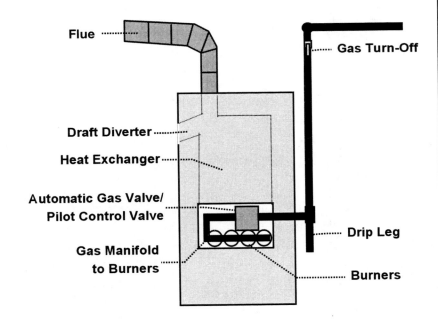

Flue · · · · · · · · · · ·

Gas Turn-Off

Draft Diverter · · · · · · · ·

Heat Exchanger · · · · · · ·

Automatic Gas Valve/
Pilot Control Valve

Drip Leg

Gas Manifold
to Burners

Burners

Gas Controls

The volume and pressure of gas delivered to the burner is controlled by a metering valve. This main control for a gas-fired furnace or boiler is normally a combination control or box-like unit housing the **main automatic gas valve** and the **pilot control valve**.

The main automatic gas valve starts and stops the flow of gas to the burners. The pilot control valve starts and stops the flow of gas to the pilot.

A pilot is a small flame that ignites gas at the burners. A standing pilot is one that burns continuously. An intermittent pilot lights only on a call for heat.

A thermocouple is a bimetallic element that senses whether or not the pilot is lighted and controls the pilot control valve, turning off the flow of gas to the pilot when the pilot flame is out.

These main controls on the gas-fired system respond to electrical signals from the **thermostat** to start gas flow to the burners when temperatures fall below a preset point and to stop gas flow when temperatures reach a desired level. The controls also respond electrically to a **limit control** on the heating unit to turn off the burners when the unit overheats due to some malfunction. The limit controls, which sense air temperature in furnaces and water temperature in boilers, turn off gas flow to the burners when high temperature limits are reached.

- **The pilot control valve and thermocouple:** Most gas-fired systems have a pilot burner, whose purpose is to light the burners when heat is called for and the main gas valve opens. A separate gas line, usually small flexible tubing, runs from the main gas control unit to the pilot burner.

Conventional gas-fired systems have a **standing pilot**, which means that it burns continuously. The standing pilot uses a safety feature called a **thermocouple** that senses whether or not the pilot is lighted. The thermocouple is an element made of 2 dissimilar metals which when heated by the pilot, completes a circuit to the pilot control valve, keeping it open and letting gas flow to the pilot. If the pilot goes out, the thermocouple cools, breaks the circuit, closes the valve, and stops the flow of gas to the pilot. When the pilot is out, the main gas valve will not open.

Customers with standing pilots often ask home inspectors if they should turn off the pilot flame during the summer to save money. In areas where condensation can be a problem, causing rust inside the heater, the pilot can provide enough warmth to avoid this problem. If condensation is not a problem, the pilot can be turned off.

Mid-efficiency gas-fired heating systems may have an **intermittent pilot**, which means that the pilot is lighted only when there is a call for heat. The intermittent pilot is ignited by a spark plug and also uses a thermocouple to verify ignition. Some mid-efficiency and high-efficiency furnaces and boilers may not have pilots at all. They use **electronic ignition** with a hot surface ignitor or electric spark which is energized by the thermostat to ignite the gas. A thermocouple-like safety device will shut down the system if the gas is not ignited.

- **Main automatic gas valve:** Normally, the main gas valve is housed in a combination unit with the pilot control valve. This valve starts and stops the flow of gas to the burners and meters the incoming gas stream for constant volume and pressure to the burner. The main gas valve receives feedback from the other control and safety devices (thermostat, limit control, thermocouple, and pilot control valve) to determine whether it is safe or appropriate to allow gas into the combustion chamber.

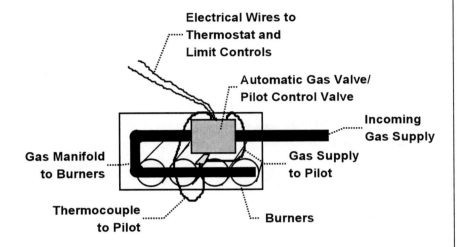

#4 Burner area in a conventional gas-fired furnace

*Photo #4 shows the **burner area in a conventional gas-fired furnace**. This unit has a standing pilot which is located between the 2 center burners below. The combination gas valve and pilot control valve is the silver-colored box-like unit. The incoming gas line approaches it from the left. The gas manifold exits from the right and moves downward and crosses in front of the burners. The thermocouple is the lighter element between the gas valve and the pilot. The other tubing between the unit and the pilot is the pilot gas supply. The electrical wiring can also be seen. The burners shown in this furnace are ribbon burners. They extend back into the furnace under the heat exchanger (not seen).*

For Beginning Inspectors

As you begin your study of heating systems, try to view as many different systems as possible. Hands-on experience is truly an important part of the learning process.
Start with conventional gas-fired furnaces and boilers. Ask your friends if you can look at their units and have them fire up the unit so you can observe the flames. Become familiar with the gas controls, the burners, pilot, thermocouple, and so on. Move on to view some of the newer mid-efficiency and high-efficiency models.

For Your Library

It pays to stop in at retail heating outlets to see new models. Pick up their pamphlets and start your own files of these heating systems.

Gas Burners

The heat exchanger in the conventional and mid-efficiency gas-fired heating system is a semi-sealed firebox, open at the bottom where the gas burners are. Burners in old gas-fired systems were star shaped or circular. The later, conventional systems have **ribbon burners** or tubes, each with a row of small holes or ports where the gas comes out.

A fuel gas, if just allowed to flow from the end of a tube, will burn with a yellow, smoky flame containing unburned gas and life-threatening combustion byproducts such as carbon monoxide (CO). But gas in a heating system is premixed with air *before* burning so the flame will be blue and the byproduct mainly carbon dioxide. This air, called **combustion or primary air**, is admitted to the burner in conventional systems through an **air shutter** located where the burner tube connects to the manifold. Set screws fit into slots on each tube and are used to adjust the shutters for the ratio of gas and air for proper flame appearance.

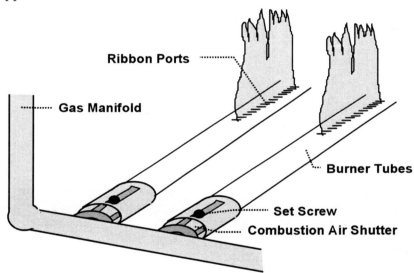

During the inspection of the gas burners, the home inspector can note the condition of the flames above the burner and determine whether the proper mix of gas and combustion air is present. Gas burners can become corroded and their ports clogged, interfering with proper burner operation. The following conditions will alert the inspector to problems:

- **Flames roaring and dancing on the burner:** This condition indicates that the flow velocity of the gas/air mixture is greater than the burning rate. It allows unburned gas to escape up the flue and creates CO due to incomplete combustion. Combustion air and/or the gas input rate needs to be adjusted.

- **Flashback or flames with a blowtorch sound burning inside the burner or near the manifold:** This condition occurs when the flow velocity of the gas/air mixture is lower than the burning rate. This problem causes soot build-up and CO due to incomplete combustion. Combustion air needs to be reduced or gas flow increased.

- **A bang or pop when the burner shuts down:** This condition may be followed by burning in the burner head. There may be too much combustion air present, a low gas flow, or a faulty burner.

- **Flames with yellow tips:** This condition is usually caused by a lack of combustion air and can be fixed by increasing its flow. However, it can indicate dirty or clogged burners or a heat exchanger problem. Soot buildup and CO production is the result of this problem.

- **Fluctuating flames:** When the flame varies greatly during burner operation, it can be caused by dirty burners or by erratic gas pressure.

- **Floating flames:** When flames are long and shapeless and seem to float above the burners, it's usually caused by a lack of combustion air. It could indicate an improper air shutter adjustment, dirty burners, or a blocked flue or chimney. The condition is dangerous because flames can roll out of the combustion chamber, causing a fire hazard. The production of CO is always a result of floating flames.

- **Flames rolling out of the combustion chamber:** Flame rollout is the end result of a floating flame condition, indicating that ignited gas is spilling out of the chamber. In addition to being a fire hazard, rollout can scorch other heating components, including wires and safety controls.

- **Flame disturbance or color change when the fan kicks on:** In furnaces, a disturbance in flames or changing flame color when the fan is in operation is an indication of a faulty heat exchanger, which is a major defect in the heating system. Flames may lift off the burner, distort, or roll out of the combustion chamber.

FLAME PATTERNS

- Dancing flames
- Flashback
- Extinction pop
- Yellow flames
- Fluctuating flames
- Floating flames
- Flame rollout
- Disturbance or color change when furnace fan kicks on

Draft air, also called secondary air, is the air required to insure the discharge of exhaust gases through the flue
.

The draft hood and draft diverter are devices that protect the heating unit from excessive updrafts and chimney downdrafts.

NOTE: In high-efficiency furnaces and boilers, the combustion chamber is sealed. In some units, there is a porthole that the home inspector can look into to verify ignition. In others, there is no visual access at all to the combustion chamber. High-efficiency units may have burners in the sealed combustion chamber, or in the case of the pulse furnace, operate by igniting an air/gas mixture released in short pulses directly from a valve.

In addition to combustion air, **draft or secondary air** is also needed to carry air through the heat exchanger and to discharge the combustion products through the flue. Draft air enters the heat exchanger through the spaces around the burners. Proper ventilation must be available to provide enough combustion and draft air to the heating unit. As mentioned on page 14, gas-fired heating units in confined spaces require 1 square inch of free vent area for every 1,000 BTU's of input.

The operation of gas burners requires a small, steady amount of draft air passing through the heat exchanger in order to insure that exhaust gases pass up the chimney, but not excessive amounts. The

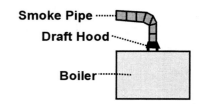

draft hood on a boiler (pictured here) and the **draft diverter** on a furnace (see illustration on page 21) protect the unit from strong updrafts by allowing basement air to be drawn into the flue directly rather than through the combustion chamber. They also protect the unit from downdrafts from the chimney, which could extinguish the pilot flame, by diverting them away from the heating unit. Air should always pass into a hood or diverter, not be flowing out of it. Any spillage from the draft hood or draft diverter during normal operation means there's a malfunction in the smoke pipe or flue, which could be blocked. Because exhaust gases can exit through the draft hood or diverter into the home, this condition represents a **safety hazard**.

Inspecting Gas Components

We're going to talk about inspecting only the gas components of a gas-fired furnace or boiler in this section. These components are similar regardless of what type of heating system they power. We'll discuss the inspection of different types of furnaces and boilers in detail later in this guide.

The inspection of the gas components on a heating unit includes checking the following:

- Meter, gas lines, and gas turn-off
- Combustion air for air flow to burners
- Flame shield condition
- Burner condition
- Burner ignition
- Flame pattern activity
- Draft hood or draft diverter operation

Follow these procedures to inspect the **gas burners** on a conventional or mid-efficiency gas-fired heating system:

1. **Turn up the thermostat** so the heating system fires. While you are inspecting the living areas of the home, you'll be examining heat registers and radiators and noting the presence of heat sources in each room. Home inspectors usually turn up the thermostat so that the heating system kicks on during this portion of the inspection.

2. **At the heating unit, turn the unit off, using the safety switch or serviceman's switch.** With the thermostat turned up, you'll now want to turn off the heating unit so you can safely gain access to the burner area.

3. **Remove the flame shield for inspection.** The burner area will have an outer access panel and/or flame shield that prevents unburned gas from rolling out of the combustion chamber during operation. The interior surface should be inspected for evidence of rollout — scorching and burning.

4. **Examine the burner heads for rust, corrosion, and clogging.** Before the heating unit is fired up, inspect the burners for any signs of rusting and corrosion. Look for mechanical damage to the burners, any displacement, or dirt on the burners that can interfere with normal operation. Note the burner heads or ports for signs of clogging. All of these conditions should be reported in your inspection report.

GAS COMPONENTS

- Meter, gas lines, and gas turn-off
- Combustion air
- Flame shield
- Gas burners
- Draft hood or draft diverter

---

*Photo #5 shows a **dismantled burner area** in a conventional gas-fired heating unit. (Of course, the home inspector does not dismantle the components as shown in the photo.) Here, 2 of the burner tubes have been removed, and a third is pulled out of position. Notice also that the gas manifold has been disconnected from the gas valve (silver box-like unit) and removed so we can see this area better. This photo shows **serious corrosion** in the burner area from poor combustion. In this case, the home had a blocked chimney that contributed to the condition, allowing condensation to build up within the combustion chamber. Upon examining the burners themselves (although not too clear in the photo), we found that the burners were corroded and their ports clogged with dirt, which also contributed to poor combustion.*

#5 Dismantled burner area

CAUTION: When you discover corrosion of the magnitude shown in Photo #5, don't use your screwdriver to try to scrape or poke at the corroded areas. You can poke right through it and make holes. Try not to damage equipment as you inspect it.

5. **Turn the unit on, using the safety switch.** After inspecting the burner components, turn the heating unit on, using the safety or serviceman's switch, so it will fire. But first, heed these cautions:

— If the gas is turned off to the heating unit and there is a **red tag** on the gas line, do not restart the burner. The red tag indicates there is a defect in the system that must be repaired by a qualified technician.

— If gas supply is turned off to the burner *and* the pilot at the manual gas valve, do not restart the system without talking to the homeowner first. The system may be shut down due to some defect that needs correction. Find out what the circumstances are and get permission from the homeowner to fire up the unit.

— If gas supply is turned off to the burners but *not* to the pilot, it should be safe to restart the unit. But ask the homeowner first why gas supply to the burners was turned off and whether it's okay to turn the supply back on.

— Remember to **have your customer stand aside** (and you too) when you re-fire the heating system. Do not stand in front of the burner area, peering into the chamber and waiting for the unit to fire. Move to the side of the heating unit, out of harm's way. Flame rollout can occur and endanger both you and your customer.

— If for any reason the pilot goes out during your inspection, follow the directions for relighting it. You should **wait 5 minutes** before relighting to give unburned gases a chance to dissipate and become diluted. This is especially important when working with an LP gas supply, since it is heavier than air and takes time to disperse. *LP gas is extremely explosive!*

When you fire up the heating unit, the gas at the burners should light with a gentle "puh" sound. Watch for any defects in the ignition such as flame rollout, loud noises such as popping and roaring, smoke, or vibration. These are all conditions that require examination and repair by qualified technicians.

• **Observe the flame pattern during operation.** Watch the flames at the burners for those conditions noted on pages 24 and 25 — yellow color, flashback, dancing, fluctuating, floating, rollout, and other distortions.

If the unit is a forced warm air furnace, monitor the flames after the fan turns on too. If flames color changes or the fan causes the flames to distort, lift off the burner, or roll out of the combustion chamber, there is a problem with the heat exchanger. (We'll discuss the inspection of the heat exchanger later in this guide.)

BURNER PROCEDURES

• Turn off heating unit.

• Remove flame shield and inspect.

• Inspect burners.

• Fire up heating unit.

• Monitor flames.

For Beginning Inspectors

If you have a gas-fired heating furnace or boiler, follow these procedures to inspect the gas burners. If not, get permission from friends to examine their burners. Be sure to follow the safety precautions we've mentioned.

Get as much hands-on experience as you can. Pay particular attention to flame patterns so that you can easily recognize defective conditions.

Photo #6 shows a *fan safety switch*. It's the white switch on the silver-colored plate in the center of the photo. Notice also that the safety or serviceman's switch is outside the furnace cabinet at the right. The burner area lies above.

Personal Note

"During one of my inspections, the pilot was out on the furnace. After checking with the owners, I got permission to light the pilot and turn on the furnace. It turns out that the furnace had a malfunctioning gas valve so that gas was released to the burners before the pilot was lighted. When I lighted the pilot, there was a huge flame rollout. Of course, my face was right down there. I burned my forehead and singed my eyebrows.
"Please be careful in situations like this one. A burned face is not a good thing to have."
Roy Newcomer

NOTE: In some furnaces, the fan will not operate if the access panel is removed. There will be a safety switch, which when pushed in by the access panel, will allow the fan to operate. You can push this switch in with your finger to get the fan to operate during your inspection.

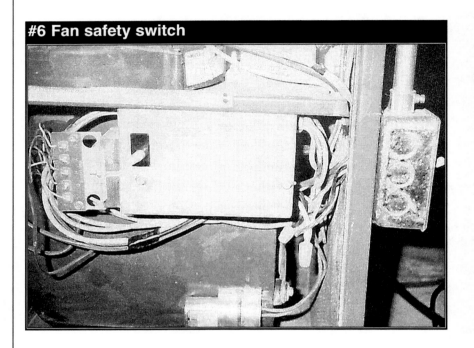

#6 Fan safety switch

While inspecting the gas components of a furnace or boiler, watch for the following conditions:

- **Gas leaks:** Any strong gas odor in the house should be dealt with immediately. Call the gas company for assistance right away and have the customer and other parties present leave the house. Don't take any chances in a serious situation, but don't push the panic button either. A faint odor of gas at pilot lights is common and isn't a leak.

However, the faint odor of gas along piping runs to the heating system is an indication of a gas leak. Be especially sensitive to gas leaks at connections at the unit itself. These conditions should be reported as a **safety hazard** with the recommendation that the proper personnel be called to correct the situation.

- **Improper LP gas tank location:** Communities have requirements for the location of LP gas tanks. Because of its explosive nature, the tanks should never be located in the house, basement, or garage. And they shouldn't be near an outside window. In general, they're required to be located at least 10' from the house. Any violations should be reported as a **safety hazard**.

- **Incorrectly sized gas meter:** Note the meter's cubic feet per hour rating and compare it to the BTU input requirements on the heating system and water heater. (Remember that a 1 cubic foot of gas is about equal to 1,000 BTU.) If the gas supply is not sufficient, this should be noted in your inspection report.

- **Improper piping or missing turn-off valve:** Follow the gas line to the heating unit and report the use of piping materials other than black iron piping. Watch for improper installation and piping support. The turn-off for the heating unit should be located near the unit.

- **Improper vent space:** For those gas-fired systems in confined spaces that use house air for combustion air and draft air, make note of whether the heating unit has enough vent space to provide the amount of air needed. Anything less than 1 square inch of free vent area for every 1,000 BTU should be reported. Vents should be present at the top and bottom of the enclosure.

- **Missing or scorched flame shield, flame rollout:** Always be sure the flame shield is still in place on a heating unit, and report if it's missing. Scorching and burn marks on the flame shield or other components near the combustion chamber should be pointed out to the customer. Explain that **flame rollout** is the cause, a condition caused when unburned gas spills out of the combustion chamber.

- **Dirty, rusted, or corroded burners:** Inspect the burners carefully before firing the furnace or boiler and report these conditions. (See page 27.) Be sure to note that the burners were not tested if the system has been shut down.

- **Evidence of CO production:** Monitor the flames for those conditions that indicate that CO is being produced as a result of incomplete combustion —flashback, yellow, floating, or dancing flames. These conditions should be reported as **safety hazards**, and you should recommend that a technician examine the heating unit and correct them. (See pages 24 and 25.) Another sign that CO is

---

CAUTION

If you detect a strong odor of gas in a home, have people leave the house immediately. Call the gas company for help.

Personal Note

"One of my inspectors ran into a situation where there was a gas leak at the furnace connection which he didn't detect before firing the furnace. Upon ignition, the flames rolled out, hit the gas pocket, and blew the furnace cover off. It flew all the way across the basement."
Roy Newcomer

- Gas leaks

- Improper LP tank location or piping

- Missing turn-off valve

- Lack of combustion air space

- Missing or scorched flame shield

- Dirty, rusted, or corroded burners

- CO production

- Draft spillage

- Improper clearances and venting

being produced during combustion is the distinctive smell of **aldehydes**, an odor like you may have smelled in a new mobile home. This smell is the result of the flame cooling when it burns against metal parts.

- **Spillage from the draft hood or diverter:** Light a match and hold it next to the hood or diverter. Move it around the area, waiting to see which way the flame leans. If the flame leans in toward the hood or diverter, that's good. But if the flame leans outward, there may be a down drafting problem and CO can be released into the home. This should be reported as a **safety hazard**. You can also use an inspection mirror to test for the drafts at the hood or diverter. A fogged up mirror indicates down drafting. The TIF 8800 combustible gas detector can also be used.

- **Improper clearances and venting:** The home inspector should examine the heating and venting equipment for the proper clearances from combustibles (see pages 13 and 14 for more information) and record any violations in your inspection report. Examine the flue for proper slope and length, holes, open joints, and corrosion. If the flue is longer than 10' or has more than one elbow, be sure that the draft is not affected. Check the opening where the flue enters the chimney flue for proper sealing with mortar. Any holes or leaks in flues should be recorded as a **safety hazard**.

The condition of the **chimney** should also be inspected. Gas-fired heating systems (except for high-efficiency units that exhaust directly to the outside) require a chimney for exhausting combustion byproducts. Type B gas vents or **metal chimneys** are also used for gas-fired units. Masonry chimneys exhausting combustion byproducts from a gas-fired heating system should have a **flue liner**, usually metal, clay, or asbestos cement pipe. Note that the home inspector is not required to inspect the interior of flues.

In general, although local codes vary, the gas burner flue should not be shared with an oil burner or fireplace. But if both a gas burner and oil burner share the same flue, the gas burner flue should enter at a point *above* the oil burner smoke pipe.

The home inspector should check for the presence of a **chimney cleanout** at the base of the chimney. There should be a cleanout leg present to catch debris from chimney and flue deterioration. Open the cleanout door carefully, as it may be full of soot and debris. Use your inspection mirror and flashlight to look up into the interior of the cleanout leg. If a great amount of debris is present, the flue may also be blocked. A blocked flue should be reported as a **safety hazard**. Recommend regular cleaning at the cleanout to prevent this unsafe condition.

NOTE: For now, we won't discuss how to report your findings about gas burners in your inspection report. This will be covered in later pages in this guide.

INSPECTION NOTE

Gas space heaters should be inspected for these same items. Always report any <u>unvented gas space heater as a safety hazard</u>. All of them produce carbon dioxide which causes headaches and other symptoms. The malfunctioning ones can produce deadly carbon monoxide.

Guide Note

The inspection of the chimney is presented in greater detail in <u>A Practical Guide to Inspecting Roofs</u>. These details of chimney inspection are not repeated here.

WORKSHEET

Test yourself on the following questions. Answers appear on page 36.

1. An 80% efficiency rating for a gas-fired heating system would best describe:

   A. A conventional system.
   B. A mid-efficiency system.
   C. A high-efficiency system.

2. A gas heating system and water heater with an input requirement of 250,000 BTU should be supplied by a gas meter rated for at least:

   A. 2.5 cubic feet per hour.
   B. 25 cubic feet per hour.
   C. 250 cubic feet per hour.
   D. 2,500 cubic feet per hour.

3. The gas manifold in a gas-fired heating system delivers gas:

   A. From the pilot control valve to the pilot.
   B. From the meter to the turn-off valve above the heating unit.
   C. From the drip leg to the gas valve.
   D. From the gas valve to the burners.

4. What is a thermocouple?

   A. A device that senses if the pilot is lighted or not and controls the flow of gas to the pilot burner.
   B. An automatic valve that starts and stops the flow of gas to the burners.
   C. A combination control containing the gas valve and pilot valve.
   D. A glow coil or electric spark that is energized by the thermostat to ignite the gas.

5. What is combustion air in regard to a gas-fired heating system?

   A. Air used to mix with gas before burning.
   B. Air used to discharge exhaust up the flue.
   C. Air that comes down the chimney.

6. What condition is usually the cause of yellow-tipped flames in a gas heating unit?

   A. Lack of draft air
   B. Lack of combustion air
   C. Gas/air mixture flow velocity greater than the burn rate
   D. Gas/air mixture flow velocity less than the burn rate

7. What flame pattern would <u>not</u> necessarily be the cause of CO production?

   A. Dancing flames
   B. Flashback
   C. Fluctuating flames
   D. Floating flames

8. What does a red tag on the gas line at the heating unit mean?

   A. The unit has passed inspection by a qualified technician.
   B. The unit has a defect that must be fixed by a qualified technician.

9. What function does the draft hood or draft diverter play on a heating unit?

   A. It protects the unit from downdrafts.
   B. It provides combustion air for the unit.
   C. It allows CO to escape into the home before passing up the flue.
   D. It delivers draft air to the unit.

10. What is the purpose of waiting 5 minutes before relighting a pilot that has gone out during the inspection?

    A. To give the unit a chance to cool down before putting your hands near the pilot.
    B. To allow gasses to dissipate and become diluted.
    C. To prevent over cycling.

Chapter Four

OIL-FIRED SYSTEMS

In this chapter, we're going to discuss the principles, components, and the inspection of the components of oil-fired heating systems, regardless of the type of heating system the burners serve. The study of particular heating systems (warm air, hot water, steam, and so on) will begin on page 46.

Oil Supply

Fuel oil is supplied in commercial grades that vary from #1 to #6 — #1 being the most refined and similar to kerosene, #6 being the least refined and heavy and dark. Home heating systems use #2 fuel oil, which is a clear, yellowish, syrupy liquid like diesel fuel that is specially blended to pump at low winter temperatures.

Fuel oil is delivered to the home and stored in a **steel oil storage tank** either buried underground outside or sitting inside the home since the oil is considered safe. It does not ignite spontaneously and will extinguish a match dipped into it. There are local codes regarding the size of inside oil tanks. In general, the inside tank has a 275 gallon capacity, although some areas allow tanks with a capacity of up to 660 gallons.

Oil tanks have a lifetime of about 20 years. As the oil level in the tank falls, air in the tank condenses, causing water to sit on the bottom of the tank. Over time, the tank begins to rust out at the bottom. With interior tanks, the home inspector should always feel along the bottom of the tank to check for leaks and rust holes. Tanks can be patched or plugged, but eventually the tank must be replaced. Underground tanks, especially those over 20 years old, should be pressure tested for leaks.

Inside tanks will have a 2" **capped fill pipe** from outside the home to the tank and a 1 1/4" **vent pipe** returning to the outside.

Guide Note

Pages 35 to 49 show the study and inspection of the oil components of oil-fired heating systems.

#7 Oil fill and vent pipes

*Photo #7 shows **oil fill and vent pipes.** Notice that the vent pipe is protected by a rain cap to keep water out of the tank. If the homeowner converts an old oil system to a gas heating system, these pipes should be removed or sealed to prevent any chance of the oil company making an unexpected delivery.*

Interior oil tanks must be located at least 5'- 10' (note local requirements) from the oil burner. There should be a **manual oil turn-off** and an **oil filter** between the tank and the oil burner. Fuel oil naturally contains small amounts of solid matter that must be filtered out of the supply. The oil filter should be changed annually. The filter may be located on the tank end of the supply line or just before the oil burner at the other end.

The **oil supply line** to the oil burner is usually 3/8" or 1/2" OD (outside diameter) copper piping, although you may occasionally see steel or galvanized piping. Often, the supply line is buried in the basement floor, going into the floor just after exiting the oil tank and resurfacing at the oil burner. If the copper piping lies on or just above the floor, it should be protected from mechanical damage, denting, or crimping. If piping is not protected or is already damaged, the home inspector should make a note of that fact in the inspection report.

Worksheet Answers (page 34)

1.	B
2.	C
3.	D
4.	A
5.	A
6.	B
7.	C
8.	B
9.	A
10.	B

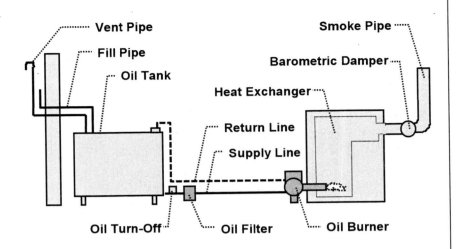

Vent Pipe ····
Fill Pipe ····
Oil Tank ····
Return Line ····
Supply Line ····
Oil Turn-Off ····
Oil Filter ····
Oil Burner ····
Smoke Pipe ····
Barometric Damper ····
Heat Exchanger ····

If oil is fed to the oil burner by gravity, there will be only one line between the tank and the burner. Any excess oil at the oil burner nozzle is fed through a bypass back into the burner. There may be 2 lines between the tank and the burner — the supply line and a **return line** to return excess oil to the tank. This is usually done if the tank is below the level of the burner or at some distance from the burner.

#8 Interior oil tank

*Photo #8 shows an **interior oil tank**. The 2 pipes at the top right of the tank are the fill and the vent pipes to the exterior. At the bottom left of the tank, you can see the supply line from the tank, the small manual turn-off valve, and the box-like oil filter. In this case, the supply line to the burner is buried under the basement floor. This setup has no return line from the burner to the tank.*

The <u>primary control</u> on an oil burner starts and stops the burner in response to thermostat signals and verifies ignition. A primary control may be located on the smoke pipe or on the burner housing.

"I remember an incident in our area where the homeowner had left the oil fill pipe in place after converting to a gas heating system. The oil company arrived by mistake and pumped 30 gallons of oil into this fellow's basement before the error was discovered!

"And I've read about a case in Minneapolis where the oil company mistakenly pumped 150 gallons into an out-of-use fill pipe, not only filling the basement but contaminating the water supply and saturating the soil in the process. The oil company had to buy the house and remedy the situation."

Roy Newcomer

Oil Controls

The main control for an oil burner is called the **primary control**. This control may be combined in a single unit with the **safety control**. Both controls together are often called the primary control, the primary safety control, the primary safety relay, and so on. We'll use the term *primary control* to mean both.

The primary control to the oil burner responds to electrical signals from the **thermostat** to start the burner motor and to energize the ignition transformer. When the burner starts, oil is sprayed into the combustion chamber, and the ignition transformer ignites the oil spray. If the flame is not established some time after ignition (15, 30, or 45 seconds), the primary control will turn off the flow of oil to the burner and the burner will shut down. Some primary controls will make a second attempt at ignition, waiting again to verify that ignition has taken place. The primary control responds again to the thermostat when temperature levels are reached to shut down the burner.

The primary control also responds electrically to the **limit control** on the heating unit to shut down the burners when the unit overheats due to some malfunction. The limit controls, which sense air temperature in furnaces and water temperature in boilers, turn off the flow of oil to the burner when high temperatures are reached.

- **Primary control on stack:** Older oil burners have the primary control located in the smoke pipe or exhaust stack on the heating unit. The stack-type primary control, sometimes called a **stack relay**, has a bimetallic coil, or temperature sensor, that extends from the back of the control into the smoke pipe. It verifies the presence of a flame in the combustion chamber by the rise of temperature of the exhaust gases in the smoke pipe. If no heat is sensed after a certain time, the burner will be shut down. The primary control has a **reset button** on it. If the oil burner is shut down after a failed ignition, the reset button can be pushed *once* to override the control and restart the burner. Pushing it more than once could allow an unsafe ccumulation of oil in the combustion chamber.

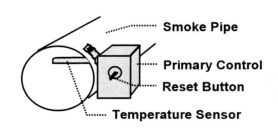

- **Primary control on burner:** Oil burners in recent years have the primary control located on the oil burner housing itself. This type of primary control has a cadmium sulfide **photocell** which sits in this assembly or on the blast tube and "sees" the flame in the combustion chamber in order to verify ignition. One brand of photocell is called a Fire Eye. If the photocell doesn't detect a flame, it will assume that oil is not being ignited and shut down the burner pump. This type of primary control also has a reset button.

Oil Burners

Modern oil burner are **gun-style burners** that force oil under pressure through a nozzle to break it into small particles. These oil particles are mixed with air to form a spray. The oil spray is ignited by an electric spark. The earlier gun burners ran at 1,750 rpm (revolutions per minute), but the new high pressure, high speed models run at 3,500 rpm.

The components of the oil burner are packaged together in a single housing, usually cast aluminum, and are as follows:

- **The primary control**, as described above, may be located on top of the burner housing.

- The **burner motor**, located at one end of the burner housing. Squeaks, grinds, and other noises in the motor are an indication that bearings are wearing out or the unit needs maintenance.

- A **squirrel cage fan** is located between the motor and the fuel pump. **Air shutters** and shutter adjustment controls are located on the fuel pump side of the fan. The fan forces air into the combustion chamber. The fan should run smoothly without excess noise that could indicate rotating parts are out of balance dirty or worn.

As with gas-fired heating systems, the oil-fired system needs an adequate amount of combustion air. Proper ventilation must be provided — 1 square inch of free vent area for every 1,000 BTU of input. If the heating unit is located in a confined area, vents should be installed at the top and bottom of a doorway into the area.

OIL BURNER COMPONENTS

- Primary control
- Burner motor
- Fan
- Fuel pump
- Ignition transformer
- Blast tube with nozzle and electrodes

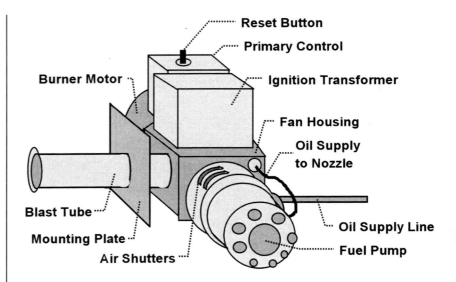

- The **fuel pump**, located at the other end of the burner housing, connects to the fuel line. Another fuel line runs from the fuel pump to the blast tube. A pump adjustment control is present on the surface of the pump. The fuel pump raises oil pressure to 100 pounds per square inch (psi) and pumps it to the nozzle in the blast tube. A **strainer** in the pump, in addition to the filter in the oil line, removes debris from the fuel so it does not clog the nozzle.

- The **ignition transformer**, supplying power to the ignition electrodes, is located on top of the oil burner housing.

- The **blast tube**, which extends into the combustion chamber is mounted on the heating unit by means of a mounting plate, and contains the **nozzle** and **ignition electrodes**. Pressurized oil from the oil pump is forced through the nozzle. The resulting oil particles are mixed with combustion air, forced into the blast tube by the fan. The mixture is ignited by an electric spark between the ignition electrodes, which are powered by the ignition transformer.

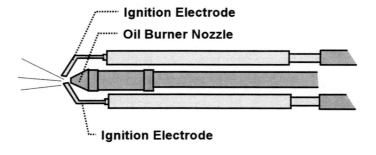

The end of the blast tube normally has a slotted **combustion head** or slotted metal piece through which the air enters the combustion chamber. Newer oil burners, called **flame retention burners**, have a cone shaped flame retention ring that gives more pressure, velocity, and rotation to the air stream, providing more efficient combustion.

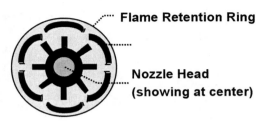

Flame Retention Ring

Nozzle Head (showing at center)

The resulting flame is spherical in shape and does not move far from the end of the blast tube. Older combustion heads cause the flame to be larger and shapeless within the chamber, therefore less efficient.

- The **combustion chamber** or firebox for older oil burners will be lined with **firebrick**, called refractory, or can have a heavy metal shield. Over time, firebrick can deteriorate to the point that the metal walls of the heating unit are exposed. Flame retention burners can be used with a combustion chamber lined with a ceramic fiber.

There should be an **inspection port** at the front of the combustion chamber from which to examine the flames. The home inspector should be familiar with the following flame conditions:

— **Bright orange flames:** An oil burner flame should be bright yellow orange in older low pressure burners to yellowish white in high pressure burners.

— **Sooty flames:** Flames that have sooty or smoky edges are starved for combustion air.

— **White flames:** A pure white flame is an indication of too much combustion air.

— **Blue or red flames:** These color flames also indicate a problem with the burner adjustments. For any circumstance where flames are not bright orange or yellowish white, a specialist should be recommended to adjust the burner or replace a faulty or clogged nozzle Unusual flame activity can indicate a cracked heat exchanger.

Definitions

The *oil burner nozzle*, located in the blast tube, shoots out oil particles into the firebox. *Ignition electrodes*, also located in the blast tube, work in pairs to provide a spark that ignites the oil.

The *flame retention burner* has a flame retention ring combustion head that provides increased air pressure, velocity, and rotation for more efficient combustion.

FLAME CONDITIONS

- Okay: bright orange, from yellowish orange to yellowish white

- Not okay: White, blue, or red flames or flames with sooty or smoky edges

Photo #9 shows an oil-fired boiler. The pump at the lower right is the boiler circulating pump. The unit at the front of the boiler (left) is the oil burner. Here, you can see the primary control in the silver-colored box with the reset button. The black box behind it is the ignition transformer, and you can see the burner motor below the primary control. The silver-colored blast tube extends into the combustion chamber. Note the large door to the combustion chamber — this was a coal boiler before converted to oil. Note the oil filter at the far left.

#9 Oil-fired boiler

The proper draft over the firebox is very important to the efficient operation of an oil burner. Most oil burners have a **barometric damper** located in the smoke pipe over the heating unit. The barometric damper is a balanced hinged plate in a frame that either remains closed or swings open when pressure inside the flue falls. The damper allows for a constant draft in the chimney when the system is on. House air is drawn into the exhaust flue, but exhaust gases are not allowed to escape out of the flue. If the damper is missing or inoperative, it should be noted in your inspection report.

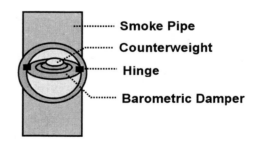

- Smoke Pipe
- Counterweight
- Hinge
- Barometric Damper

Definitions

A barometric damper, located in the smoke pipe above an oil-fired heating system, is a hinged plate that swings open or closed to regulate drafts.

An electronic vent damper automatically opens before the burner starts and closes when exhaust gases cool down.

The bimetallic vent damper expands open when heated and contracts closed when cooled.

#10 Barometric damper

*Photo #10 shows a **barometric damper** in the smoke pipe above an oil furnace. It's the circular damper located at the bottom of the photo, just above the furnace. This setup also has an **electronic damper** directly above the barometric damper. This damper opens about 30 seconds before the burner is allowed to fire and closes if the exhaust gases are too cool. If this type of damper malfunctions in the closed position, it will prevent exhaust gases, including CO, from escaping and let them enter the house. That poses a serious safety hazard.*

Another kind of vent damper found is the **bimetallic damper**, which is not electrically driven. It will expand and open when it becomes heated and will contract or close when it becomes cool. We find that these dampers often fail and recommend they be removed.

Inspecting Oil Components

We're going to talk about inspecting only the oil components of an oil-fired boiler or furnace in this section. These components are similar regardless of what type of heating system they power. We'll discuss the inspection of different types of furnaces and boilers in greater detail later in this guide.

Follow these procedures to inspect the operation of the oil burner:

1. **Turn up the thermostat** so the heating system fires. Continue with the inspection of the home's living spaces, noting the presence of heat sources and so on.

2. **At the heating unit, turn the unit off, using the safety switch or serviceman's switch.** With the thermostat still turned up, turn off the heating unit so you can examine the combustion chamber.

BURNER PROCEDURES

- Turn off heating unit.
- Open porthole and inspect firebox.
- Fire up the heating unit.
- Observe ignition.
- Monitor flames.
- Inspect front of unit.
- Listen to oil burner.

#11 Oil furnace

Photo #11 shows an *oil furnace*. Notice the round porthole above the oil burner.

For Beginning Inspectors

If you have an oil-fired furnace or boiler, follow these procedures to inspect the oil components. If not, find a friend with one and conduct your examination. Be sure to follow the safety precautions we've mentioned.

3. **Open the porthole to the combustion chamber and inspect the chamber lining.** With the system turned off, open this porthole and use your flashlight to inspect the interior of the combustion chamber for firebrick deterioration or collapse. A badly adjusted oil burner can shoot the oil spray against the far wall of the chamber. This condition cools the oil so it doesn't burn clean and further erodes the firebrick.

NOTE: In some cases, you may find an oil-fired system with a porthole partially open as an adjustment to the burner. Sometimes, the porthole may be sealed shut. Don't mess with either one. If you can't examine the chamber or view the flame, record that fact in your inspection report.

4. **Close the porthole and turn the unit on, using the safety switch.** After inspecting the combustion chamber, fire up the system, heeding these cautions:

— If the oil supply is turned off to the heating unit, do not start the system without talking to the homeowner first.

The heating unit may be shut down due to some defect that needs correction. Find out what the circumstances are and get permission from the homeowner to fire up the unit.

— If the heating system was in operation before you turned it off and it fails to ignite on restart, you can push the reset button on the primary control *once*. If the unit still won't ignite, don't push the reset button any more as you could cause oil to accumulate in the combustion chamber.

— **Remember to have your customer stand aside** (and you too) when you re-fire the heating system. Faulty or late ignition can cause too much fuel to ignite at once, resulting in a **puffback** of flame, smoke, and soot from the combustion chamber.

5. **Observe ignition.** When you fire the system, watch for any defects such as puffbacks, loud bangs, oil smells, or soot that accompanies ignition. These are signs of inadequate maintenance, poor burner adjustments, or some other defect in the burner.

6. **Observe the flames.** The porthole should have an eyeglass in it, through which you can look to observe the flames in the combustion chamber. Take another look at **Photo #11** to see the eyeglass in the porthole. Watch for flame conditions as noted on page 41.

7. **Examine the front of the heating unit.** While the heating unit is in operation, check out the front of its cabinet. Notice any evidence of soot on the front near the porthole and on the burner. **Photo #11** shows soot accumulation below the porthole, indicating that the flame had been kicking out of the chamber. Look for any cracks or open joints around the blast tube, mounting plate, and porthole. If you can see flames through cracks and joints, point this out to your customer and recommend sealing them with refractory caulk or cement.

8. **Listen to the oil burner:** There's little the home inspector can do to inspect the oil burner itself. But you can listen to it and report noises such as squeaking or grinding that indicate malfunction. Excess vibrations may mean that mountings are loose.

Personal Note

"Oil-fired systems can produce a lot of dirty stuff. Here's an interesting story to show just how much soot was hiding in such a system.

"One of our instructors at the American Home Inspectors Training Institute has friends whose oil furnace exploded. So much soot blew out. They not only had to replace the furnace and have all the ductwork cleaned, but had to have the furniture professionally cleaned before the place was back to normal."

Roy Newcomer

Personal Note

"I once opened a chimney cleanout door to find the cleanout plugged with soot. The owner asked if I would clean it out for her. Well, I agreed to do it.
"I put a box under the cleanout door and gave the soot a poke. Stuff kept coming out and coming out. It overflowed the box and kept coming. Soot went all over the place. I wish I hadn't agreed to do it."

Roy Newcomer

In general, while inspecting the oil components of a furnace or boiler, the home inspector should watch out for the following conditions:

- **Leaking oil tank:** Feel along the bottom of the interior oil tank for leaks. A leaking tank should be replaced or repaired. If the property has an underground tank, be sure to let your customers know about it. Some local codes require a leak test for an underground tank before the home can be sold.

- **Improper piping or missing turn-off valve or filter:** Follow the oil line from the tank to the oil burner and report if the oil piping is above the floor and subject to mechanical damage. Copper piping should be protected. Locate the turn-off valve and the oil filter. If either one is missing, be sure to report the condition.

- **Improper combustion air space:** For those oil-fired systems in confined spaces, make note of whether the heating unit has enough vent space to provide sufficient combustion air.

- **Oily smoke smells or soot:** An oil burner should operate odor free. Any noticeable odor of oily smoke in the area of the heating unit indicates that the oil burner requires adjustment or repair. A properly functioning oil burner should not produce smoke at the chimney nor should it produce soot. Check for soot at the top of the chimney, at the firebox porthole, at the barometric damper, or in the area near the furnace or boiler. A specialist should be called in to examine the oil burner.

- **Deteriorating firebrick or firebox leakage:** Let customers know if you find cracks, broken sections, or holes in the firebox lining. This can be a **safety hazard**, especially if the flames are burning through the firebox walls. Check for cracks and open joints at the firebox. Recommend that such openings be sealed.

- **White, red, blue, or sooty flames:** Anything other than bright orange or yellowish white flames in the firebox are indications of inefficient combustion. They are an indication of an oil burner out of adjustment, a clogged nozzle, or other malfunction. Customers should be reminded that an oil burner must be **serviced annually**.

- **Noisy or vibrating oil burner:** These are signs of needed maintenance and should be noted in the inspection report.

- **Spillage from the barometric damper:** An oil burner typically produces around 15% carbon monoxide (CO), so spillage of exhaust air at the barometric damper is especially dangerous and should be considered a **safety hazard**. Report if the damper is missing, broken, or malfunctioning in any way. When the heating unit is in operation, light a match and hold it next to the barometric damper opening in the smoke pipe. If the flame leans in toward the smoke pipe, exhaust is moving up the chimney. If the flame leans outward, it's an indication of a down drafting problem and CO is being released into the home.

- **Improper clearances and venting:** The home inspector should examine the heating and venting equipment for the proper clearances from combustibles (see pages 13 & 14 for more information) and record any violations in the inspection report. Examine the smoke pipe for proper slope and length, open joints, and corrosion. If the smoke pipe is longer than 10' or has more than one elbow, be sure that the draft is not affected. With oil-fired systems, it's especially important to check the junction of smoke pipe and chimney to be sure it's tightly sealed with mortar. Any holes or leaks in the venting systems should be reported as a **safety hazard**.

The condition of the chimney should also be inspected. Oil-fired heating systems require a chimney for exhausting combustion byproducts and should have a **flue liner**. The oil burner should not share a flue with a gas burner or a fireplace, although local codes may allow it. If both an oil burner and gas burner share the same flue, the oil burner smoke pipe should enter the flue at a point *below* the flue for a gas burner.

INSPECTING OIL COMPONENTS

- Leaking oil tank
- Improper or unprotected piping
- Missing turn-off or oil filter
- Lack of combustion air space
- Oily smoke smells or soot
- Deteriorating firebrick or leaking firebox
- Wrong-colored flames
- Noisy or vibrating burner
- Draft spillage
- Improper clearances and venting

#12 Improper smoke pipe

*Take a look at **Photo #12** which shows an **improper smoke pipe**. This pipe has more elbows (bends) than allowed by local code. It's also too long — so long that the combustion gases cool by the time they reach the chimney. This smoke pipe is also covered with corrosion.*

The home inspector should inspect the **chimney cleanout** at the base of the chimney as described on page 33.

#13 Open cleanout door

*Photo #13 shows an **open cleanout door** at the base of the chimney. Look into the cleanout with flashlight and mirror to check for flue blockage.*

For now, we won't discuss how to record your findings about oil burners in your inspection report. This will be covered in later pages in this guide.

NOTE: The **pot furnace** is another type of oil-fired heating system. It's no longer popular, but may still be found. They are usually located within the living space and furnish heat by convection. The pot furnace consists of a pot containing a pool of oil. The heat of the pot causes the oil to vaporize and, when mixed with air, can be ignited. For safe operation, the pot furnace must be vented through a chimney. The home inspector won't run across these very often. If you do, be sure the unit is vented.

WORKSHEET

Test yourself on the following questions.
Answers appear on page 52.

1. What causes oil tanks to rust?

   A. The vent pipe allows debris into the tank.
   B. The air in the tank condenses and forms water.
   C. Oil is naturally corrosive.
   D. The tanks aren't manufactured to last.

2. Where will the oil filter be located in the oil supply line?

   A. On the exterior fill pipe
   B. At the top of the oil tank
   C. Between the tank and the oil burner
   D. In the heating unit

3. What is <u>not</u> a function of the primary control on an oil burner?

   A. To sense air temperatures in furnaces
   B. To start and stop the burner motor
   C. To energize the ignition transformer
   D. To prove ignition

4. Name the 6 main components of an oil burner as shown here.

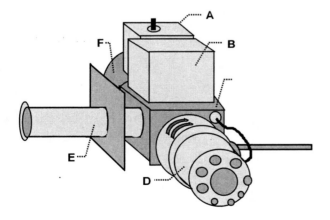

5. The oil burner primary control may be located on the smoke pipe above the heating unit.

   A. True
   B. False

6. What type of oil burner has the greatest operating efficiency?

   A. A pot furnace
   B. A low pressure gun-style burner
   C. A high pressure gun-style burner
   D. A flame retention burner

7. What condition is indicated by bright orange flames in the firebox of a oil-fired system?

   A. Flames are starved for combustion air.
   B. There's too much combustion air.
   C. The oil burner is properly adjusted.
   D. The nozzle is probably clogged.

8. What function does a barometric damper play on an oil-fired heating unit?

   A. It supplies power to the ignition electrodes.
   B. It regulates drafts to and from the heating unit.
   C. It mixes combustion air with oil in the burner blast tube.
   D. It allows CO to escape into the house.

9. What should the home inspector do if the firebox porthole is sealed?

   A. Don't break the seal, then report that the firebox wasn't inspected.
   B. Break the seal, then inspect the firebox..
   C. Report the condition as a safety hazard.
   D. Push the reset button.

10. For oil-fired heating units, what clearance is required in front of the furnace or boiler?

    A. 6"
    B. 10"
    C. 18"
    D. 24"

Chapter Five

GRAVITY WARM AIR FURNACES

Some people use the term *furnace* indiscriminately when talking about any kind of heating system. However, the term *furnace* is used to mean a heating system that heats the home with warm air; the term *boiler* is used when referring to a heating system that heats the home with hot water or steam.

Operation and Distribution

When heated, air expands and becomes lighter so that it rises. When cooled, air contracts, becomes heavier, and falls. That's the simple principle behind the **gravity warm air furnace**. Air heated in the furnace rises naturally, carrying warm air to the living areas of the house, and then falls back to the furnace as it cools. This process creates a pattern of air circulation. The gravity warm air furnace has no moving parts. There are no blowers or motors to force the air into the house. The only electrical connections necessary would be for the thermostat and burner controls — gas valve (if gas-fired) or oil motor and pump (if oil-fired).

The oldest type of gravity warm air furnace was mounted in a square opening in the floor in the center of the house. Heat rose from a circular opening over a heat exchanger dome. Cool air dropped back to the furnace through the space between the square and circle. The home inspector may find these earliest gravity furnaces still in use, but converted from coal to gas or oil.

The familiar and later type of gravity warm air furnace was called the **octopus furnace**, because of the many large supply ducts extending from the top and even larger return ducts at the bottom. You may find this type of furnace in older homes, probably converted to gas or oil. They haven't been installed for over well over 60 years. These are the furnace components:

- **Jacket:** Old octopus furnaces are made from heavy-gauge iron or steel. Usually located in the center of the basement, this unit is large and takes up a great deal of space. The jacket can become scorched, burned, or rusted over time, especially if there is a spillage or back draft problem. But generally, the construction can last for years.

Guide Note

Pages 51 to 56 present the study and inspecting of gravity warm air furnaces

- **Fuel burner:** Located at the bottom of the furnace, the combustion chamber on an existing octopus furnace is now most likely modified to accommodate a gas or oil burner.

#14 Gravity warm air furnace converted to gas

*Photo #14 shows a **gravity warm air furnace converted to gas**. You can see the gas burner at the bottom of the furnace. These furnaces are extremely inefficient to run and are terrible fuel wasters. The home inspector should always inform customers of the cost of continuing to run an old heating system like this.*

Worksheet Answers (page 50)
1. B
2. C
3. A
4. A is the primary control.
   B is the ignition transformer.
   C is the fan or fan housing.
   D is the oil pump.
   E is the blast tube.
   F is the burner motor.
5. A
6. D
7. C
8. B
9. A
10. D

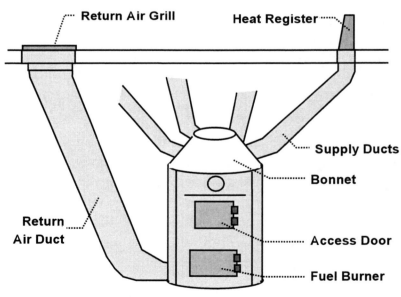

- **Heat exchanger:** A heat exchanger is a heavy metal hood above the combustion chamber that holds and contains the burner flame. In a furnace, it separates exhaust air from the circulating air that heats the house. The hot gas on one side never comes into direct contact with the circulated air. If a heat exchanger cracks or rusts through, combustion products would escape through the holes into the house's air supply.

The heat exchanger in an octopus furnace is made of seamed cast iron sections or rings and should be able to be seen and inspected by opening the front access door. The home inspector should carefully inspect the heat exchanger with flashlight and inspection mirror for cracks at seams, rusting, and deterioration. If the heat exchanger fails in an octopus furnace, the whole furnace would have to be replaced.

- **Bonnet:** The top of the octopus furnace is called the bonnet. Often, this bonnet is covered with plaster reinforced with **asbestos.** Take another look at **Photo #14** which shows an asbestos-containing covering on the bonnet. The home inspector can recommend that this area can be painted over or encapsulated to keep the asbestos in place (it's dangerous when asbestos particles are released in the air). Or asbestos can be removed. However, the cost of removal probably exceeds the cost of an entirely new furnace.

- **Supply ducts:** These ducts carry the flow of heated air upward into registers on interior walls or in the floor. Heat distribution in a house heated by a gravity air system is likely to be slow to the far corners of the house. There may be **balancing dampers** in the ducts to balance the heat supplied to various areas. If there are no dampers, the shortest ducts will get the greatest amounts of warm air.

Supply ducts may be wrapped with asbestos, usually looking like white or gray paper and held in place with metal straps at the joints. This insulation can be removed or left in place if in good condition.

- **Return ducts:** Cool air is returned to the furnace when it is drawn into the return air ducts by a natural draft resulting from the air circulation. Return air enters the ducts through grills, usually installed in the floor. The home inspector should check the living areas to be sure that return air grills have not been carpeted over. The large return ducts enter the furnace at the bottom. Note that the octopus furnace **doesn't have an air filter**. The returning air has so little force that even an air filter could stop its flow.

Definition

A heat exchanger is a heavy metal hood above the combustion chamber that holds and contains the burner flame. In a furnace, it separates exhaust air from the circulating air that heats the house.

GRAVITY WARM AIR COMPONENTS

- Cast iron or steel jacket
- Fuel burner, probably converted from coal
- Heat exchanger
- Bonnet, probably asbestos covered
- Supply ducts, maybe asbestos wrapped
- Return ducts
- No moving parts, no air filter

For Beginning Inspectors

It would be helpful if you can locate an old gravity warm air furnace for a practice inspection. Your best bet would be to try older homes in your area that may still have the original installation.

You may occasionally find an octopus furnace without return ducts. The return panel at the base of the furnace is left open and cool air, falling through the grills, drops to the basement and is pulled back into the furnace. This is a dangerous situation. If there is a malfunction or puffback from the fuel burner, combustion air can be drawn into the air return. Open returns in the basement should be reported as a major defect.

- **Smoke pipe:** The smoke pipe on the octopus furnace extends from the back of the furnace body to the chimney.

The old gravity warm air furnace can be **updated** to a forced warm air system by installing a blower unit with a filter, fan, and motor assembly. The blower unit is typically installed in the return air duct at the base of the furnace, although it may be found on top of the furnace.

*Photo #15 shows a **blower unit added to the return air duct**. By the way, notice how large the return air duct is. Even with this improvement, the old furnace is still inefficient. Replacing such a unit with a new furnace would probably pay for itself in a few years in fuel savings.*

#15 Blower unit added to the return air duct

Inspecting Gravity Warm Air Furnaces

When inspecting the gravity warm air furnace, the home inspector will be inspecting:

- The **furnace** itself including the fuel burner
- The **distribution ductwork**, registers, and returns
- The **venting system** including the smoke pipe, chimney, and cleanout.

The procedures for inspecting the fuel burner, smoke pipe, chimney, and cleanout were presented in earlier pages in this guide (see pages 24 to 31 for gas burners and pages 39 to 43 for oil burners). We won't repeat these procedures here.

The home inspector should also inspect the gravity warm air furnace for the following conditions:

- **Lack of heat source in each room:** As the home inspector examines the living spaces of the home, each room should be examined for the presence of a heat source. If no register is present, this should be noted in the inspection report. Turn up the thermostat as you inspect the living area. Holding a tissue in front of the heat register can give you an idea of the air circulation from the furnace. A tissue held against a return grill can test the pull of air back to the furnace — the tissue should be sucked against the grill.

- **Dirty, loose, or missing registers and grills:** Dust and dirt on registers and grills may be from bad housekeeping or from dirty supply ducts since the furnace does not have an air filter. But soot and dark staining on the walls above the registers can indicate a problem with a cracked heat exchanger (especially with an oil burner) that's allowing combustion air back into the air supply. This condition should be reported as a **safety hazard**. Also, watch for grills that have been carpeted over that block cool air return.

- **Defects in furnace jacket:** Inspect the furnace jacket for any holes, corrosion, or rust that may allow combustion air to escape into the house. This is especially dangerous with an oil burner. Such holes can be sealed as long as the cause of the problem is dealt with. Scorch or burn marks can indicate areas of possible failure. Staining and soot around the burner would indicate spillage and downdrafts.

- **Cracked, rusted, or deteriorated heat exchanger:** After turning the furnace off, open the front access door of the

---

INSPECTING GRAVITY WARM AIR FURNACES

- Heat source in each room
- Dirty, loose, missing, covered registers and grills
- Jacket with holes or corrosion
- Cracked heat exchanger
- Presence of asbestos
- Leaking, damaged, loose, or corroded ducts
- Open return

"When I come across an old octopus furnace, I take my time explaining to customers how it works and its level of efficiency. If the unit has no balancing dampers or blower unit added, I know they're not going to be happy with it. And they're going to pay too much for fuel. Asbestos always causes concern too. So it's best to be honest with them."

Roy Newcomer

octopus furnace and use your flashlight and inspection mirror to inspect the heat exchanger. If you discover any flaw in the heat exchanger, report it as a **major defect** and explain the danger of CO escaping into the home from this condition. With the gravity warm air furnace, the only solution to a cracked heat exchanger is furnace replacement.

The visual examination of the heat exchanger does not allow you to see 100% of it. We recommend that you always point this out to the customer and recommend that the unit be **evaluated by a furnace technician before settlement** on the home. This should be marked in your inspection report.

Remember that unusual or unstable flame action at the burner, flame rollout and puffbacks, and excessive corrosion around the burner area are indications of a faulty heat exchanger. And, as mentioned on the previous page, soot and staining around the heat registers can also mean the heat exchanger is in bad shape.

- **Presence of asbestos:** Examine the bonnet for plaster reinforced with asbestos and supply ducts for the presence of asbestos insulation. Explain the options to the customer and note the presence of asbestos-like materials in the inspection report.

- **Leaking, damaged, loose, or corroded ducts:** Inspect the condition of the supply and return ducts and their supports. Since they're likely to be old, be sure to look for open joints. Be careful about jiggling these ducts. If you cause them to open or fall, you're likely to release a lot of dirt into the basement. Check the supply ducts for balancing dampers. Inform your customer about their presence or absence. Gravity warm air furnaces without balancing dampers can be very slow to heat the far reaches of the house.

- **Open return in basement:** The situation where there are no return ducts from the living space back to the furnace can be dangerous. The open return at the base of the furnace can pull CO from a poorly adjusted burner into the home's air supply. In fact, modern local codes now prohibit this setup. With the octopus furnace, the lack of return air ducts and the open return constitute a **major defect**. A return air duct system would have to be installed.

Chapter Six

FORCED WARM AIR FURNACES

This chapter will present forced warm air furnaces, their components, and procedures for inspecting them.

Types of Forced Warm Air Furnaces

There are many types and brands of forced warm air furnaces, and each manufacturer has its own unique design and layout. However, forced warm air furnaces have basic components in common. The **conventional gas or oil furnace** contains these 3 sections typically arranged in this way:

- A **blower unit** containing fan, fan motor, and air filter at the bottom through which cool return air from the home passes.

- The **burner area** and combustion chamber in the middle where heat is produced.

- The **heat exchanger** at the top from which exhaust gases are vented and surrounding heated air is distributed.

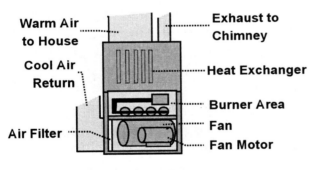

Conventional Furnace

Removing the front access panels on a conventional furnace gives the home inspector a view of the blower and burner areas for inspection. Using a flashlight and mirror, it's possible to examine a portion of the heat exchanger from the burner area, although no heat exchanger is 100% visible. In conventional gas-fired furnaces, only **about 25%** of the heat exchanger is visible for inspection; in some oil-fired furnaces, you may not be able to see the heat exchanger at all.

The home inspector may find some conventional forced warm air furnaces with an auxiliary electric heater, called an

Guide Note

Pages 57 to 76 present the study and inspection of forced warm air furnaces. It also discusses how to report your findings on furnace inspection.

Guide Note

Refer to pages 19 and 20 for more discussion on the subject of conventional, mid-efficiency, and high-efficiency furnaces.

TYPES OF FURNACES

- Conventional gas or oil
- Electric furnaces
- Mid-efficiencies gas or oil
- High-efficiency gas or oil condensing furnaces and gas pulse furnaces

Definitions

The plenum is the first large section of supply duct directly over a furnace from which smaller ducts branch out to distribute heat to the house.

An electric plenum heater is an auxiliary heater, usually added to an oil furnace, that is located in the plenum.

electric plenum heater, installed in the **plenum**, which is the first section of supply duct directly over a furnace from which smaller ducts branch out. It's most commonly used with a conventional oil-fired furnace. This auxiliary heater doesn't work at the same time as the furnace. It will first try to satisfy the call for heat from the thermostat. When it can't keep up with the heating demand, the plenum heater will switch off. Only then will the furnace kick on.

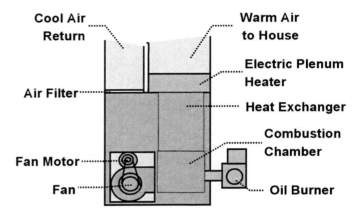

Oil Burner with Electric Plenum Heater

Some forced warm air furnaces are powered by electricity, which is by far the most expensive to operate. In **electric furnaces**, there is no combustion. Therefore, there's no need for a burner, heat exchanger, or means of exhausting combustion byproducts. These furnaces have a bank of electric resistance heaters sitting directly in the air stream. The number and size of the heating elements will vary depending on the needed heat output. Air is forced across the heating elements by the fan. Electric furnaces will typically have an electronic air filter. The inspection of the electric furnace is quite a bit different from inspecting an oil or gas furnace. Other than inspecting the filter and fan, the inspection consists largely of examining the electric furnace's operation and the general condition of the electrical wiring.

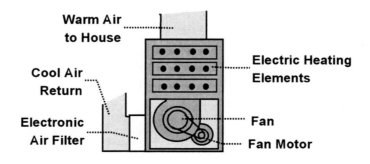

Most **mid-efficiency oil and gas furnaces** are similar in configuration to the conventional furnaces. They differ with the addition of the **induced draft fan** on the exhaust side, which is designed to pull combustion byproducts through the unit and thus reduce heat loss. Gas-fired units have an **intermittent pilot** which lights only on a call for heat. **Motorized vent dampers**, which close when the unit is not operating to prevent warm air from escaping, may be found on the smoke pipe. Some of these units exhaust hot gases through a stainless steel flue that passes through the house wall rather than the conventional chimney, making the inspection of proper clearances from combustibles very important.

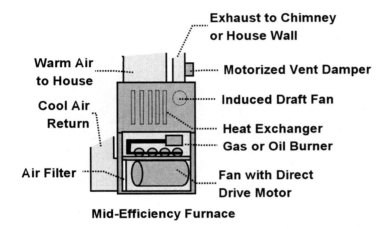

Mid-Efficiency Furnace

High-efficiency furnaces, the latest generation of furnaces, vary considerably in configuration from manufacturer to manufacturer. The combustion chamber may be located at the top or middle of the unit, heat exchangers may be located high or low and be totally inaccessible, and other components can be located in different places. In most, there is no visible access to burners. Both oil-fired and gas-fired high-efficiencies are available, although the home inspector may see far more gas units than oil.

A high-efficiency or **condensing furnace** will have 2 or even 3 heat exchangers to extract as much heat as possible from the exhaust gases, condensing a great portion of the gases. The condensate is discharged to the floor drain, while the remaining cool gases are forced with an induced draft fan through 2" PVC piping to the outside through the house wall. High-efficiency furnaces pipe in outside air to use for combustion air.

CONVERSIONS

The home inspector will find forced air furnaces converted from old gravity warm air systems or oil furnaces converted to gas. All aspects of the old setup must be compatible with the new heating system.

Definitions

An induced draft fan in a furnace pulls combustion byproducts through the unit, ensuring a good draft and reducing heat loss.

A motorized vent damper, located in the smoke pipe above the furnace, automatically opens and closes to prevent heat loss up the chimney.

Photo #16 shows a ***high-efficiency gas furnace***. *In this unit, the induced draft fan is located below the burners. Note the PVC piping at the left that exhausts cool gases to the outside through the house wall. This unit has some similarity in layout to conventional and mid-efficiencies, although other high-efficiency furnaces don't look anything like this. Often, the combustion chamber is completely sealed so the burner is not visible. Such units may have a porthole on the combustion chamber so the inspector can look into the chamber and verify combustion, but others have no visual access at all. High-efficiency furnaces can be difficult to inspect because very little of it may be visible. In general, the home inspector can check their operation and controls, general condition, and that's all.*

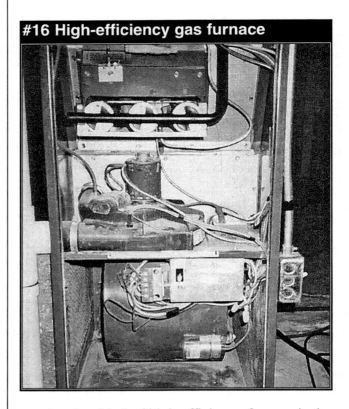

#16 High-efficiency gas furnace

Personal Note

"Because new furnaces continue to come on the market, it's difficult to keep up with every model. You simply can't expect to know everything about each one before you become a home inspector. You'll continue to learn on the job.
"What you can do now is stop in at retail furnace outlets and view as many of the new models as possible. Make a pest of yourself and ask to have access panels removed so you can see what's inside."

Roy Newcomer

Another kind of high-efficiency furnace is the **gas pulse furnace**. This model has an entirely different combustion process. Outside combustion air is mixed with gas in the sealed combustion chamber and ignited by spark plugs, causing the first pulse. The pulse travels down a tailpipe, hits the end of the tailpipe, and sends a shockwave back to the chamber, igniting the next pulse. This begins a self-perpetuating process of shockwaves with 60 to 70 explosions per second, allowing the blower and spark ignition to be

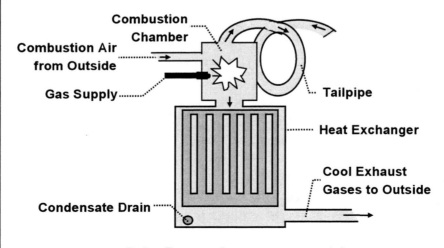

Pulse Furnace (some components)

turned off. Hot gases are forced into the heat exchanger where heat is transferred. As with the condensing furnaces, condensate is drained to the floor drain and the remaining cooled exhaust gases are vented to the outside through PVC piping.

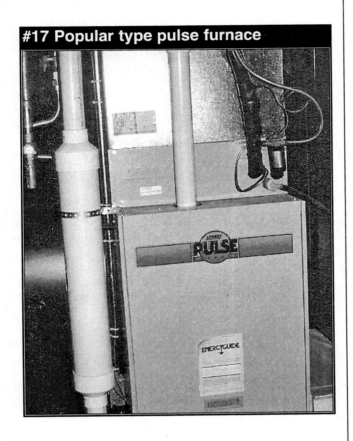

#17 Popular type pulse furnace

*Photo #17 shows a **popular type pulse furnace**. In this Lennox model, the air intake from outside is the white plastic piping at the left. Pulse furnaces can be fairly noisy. Notice the muffler on the intake pipe. Without the muffler, this furnace sounds like a train. We also found pads underneath the furnace because of the excessive vibrations. The white PVC piping at the top of the furnace jacket is the exhaust flue to the outside where low temperature combustion gases are exhausted. The condensate drain is not visible in this photo. The unit at the right of the furnace is an electronic air filter. You can't inspect the combustion chamber or heat exchanger in this furnace.*

Some forced air systems use **solar energy** as an energy source. Active systems use collectors to concentrate energy and fans to move the energy into storage and inject it into the house. The home inspector can not be expected to evaluate the performance of solar heating. But components can be inspected for condition — collectors and fans securely fastened and free of deterioration.

NOTE: Central heating that uses a **heat pump** is basically the same as a forced hot air furnace with one exception — the heating element is not a gas or oil burner, but it's a component of a reverse-cycle air conditioning system. We'll discuss the heat pump later in this guide, beginning on page 133.

FORCED WARM AIR COMPONENTS

- Furnace controls including fan control and limit control

- Blower unit consisting of a fan and fan motor

- Air filter

- Burner and combustion chamber

- Heat exchanger

- Humidifier (optional)

- Supply and return ducting

- Exhaust system

Definitions

The fan control is a furnace control that turns the fan on and off at preset temperatures. The limit control is another furnace control that turns off the burners if the furnace reaches a preset high temperature. Usually, the fan and limit control are combined in a single switch called the fan/limit switch.

Furnace Components

Forced warm air furnaces share the following components:

- **Furnace controls:** The three main controls on a forced warm air furnace are the thermostat, the fan control, and the high-temperature limit control. The **thermostat** was discussed in detail earlier in this guide (see pages 9 to 11), and we won't repeat that information here. In forced warm air systems, the thermostat turns the burner on and off as a result of a call for and the delivery of heat.

- The **fan control** is a temperature-sensitive switch that turns the furnace fan on and off at preset temperatures. The fan won't turn on until air in the plenum, heated by the burner, reaches a certain point (about 130° to 150°). After the burner turns off, the fan continues running until the air temperature in the plenum falls to the low set point (about 90° to 100°). This ensures that only warm air will be circulated to the house. When the thermostat calls for heat, only the burners should fire. If the fan begins to operate at the same time the burners fire, the fan control is in need of adjustment.

The **limit control** is also a temperature-sensitive switch that will turn off the burner if temperatures in the heat exchanger get too hot due to some malfunction in the furnace. The high temperature limit is usually set at about 170° to 200°. Typically, the limit control also functions as the fan control in a single switch, called the **fan/limit switch**.

NOTE: Some furnaces have a manual **summer switch** that allows the fan to operate independently of the burners. It may be located on the thermostat, mounted next to the safety switch, or housed on the fan control. Homeowners can turn the fan on for air circulation during the summer.

- **The blower unit:** All forced warm air furnaces have a blower unit containing a **circulating fan and fan motor**. Its purpose is to draw return air back into the furnace, push it through the heat exchanger area, and send it out through the supply ducts. Most circulating fans are squirrel cage impellers, where fan blades are on the interior surface of a cylinder that spins in its casing. With a **direct drive fan**, the fan motor is mounted within the fan casing (see photos on page 60 & 63). Older systems have the motor mounted outside the fan casing and drive the fan with a pair of pulleys and a belt (see illustrations on page 58).

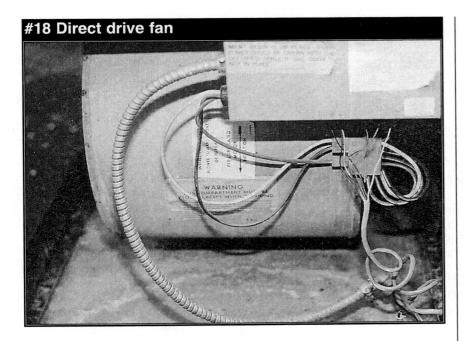

#18 Direct drive fan

*Photo #18 shows a **direct drive fan** in a gas forced air furnace. The home inspector should listen to the fan in operation for wobbling and squealing, which can be a sign of loose mountings or worn bearings. The inside of the squirrel cage fan should be checked to see if it's clean and no blades are broken or missing.*

- **An air filter:** The purpose of the air filter is to remove dust and dirt from the air before it passes into the furnace where it can clog heat exchanger passages, ducts, and registers. **Conventional air filters** are made of loose mats of glass fibers. They're mounted between the return air duct where it enters the furnace jacket and the fan. When inspecting the conventional filter, be sure that it's securely in place, facing the right way for air flow, and clean. Filters should be cleaned or changed, depending on the type, regularly every month or so. Stains around registers can be an indication of a defective filter as well as other furnace problems such as poorly adjusted burners or a cracked heat exchanger.

  There's also the **electronic air filter**, which is a separate unit, usually sitting next to the furnace at the return air duct. The electronic or electrostatic filter contains a series of fine wire grids that are given an electrostatic charge, which aids the filter in capturing small particles. When the filter is working, it pops and crackles like a bug zapper. Most standards state that home inspectors don't inspect these filters because of the danger of shock. However, we suggest that you examine the filter, but do it carefully.

- **Burner area and combustion chamber:** See pages 24 to 31 of this guide for information on gas burners and pages 39 to 43 for information on oil burners. Electric furnaces, of course, will not have a burner or combustion area. Instead, you'll find a bank of heating elements as shown on page 58.

Personal Note

"Try not to show surprise by the ignorance of some homeowners. I had to keep a straight face while teaching a home maintenance class. When I commented on the furnace air filter, one homeowner said, 'Furnaces have filters?' She'd lived in her current home for 14 years and had never once cleaned or changed the air filter."

Roy Newcomer

- **The heat exchanger:** This is the most critical component of a forced air heating system. The heat exchanger in modern furnaces is made from heavy sheet metal. It sits above and is open to the combustion chamber and burner, transferring heat from exhaust air to the air that circulates through the house. The hot gas inside the heat exchanger never comes into contact with the circulated air.

If a heat exchanger cracks or rusts through, combustion byproducts would escape through the holes into the home's air supply. **Cracks** can occur at sharp corners and at welded seams due to metal fatigue, and once started will grow because of the continual expansion and contraction of the metal. **Corrosion and rust** from too much condensation or water in the furnace can cause holes in the heat exchanger. If the burner and combustion area is rusted, corroded, and flaking, chances are the heat exchanger is too. Because faulty heat exchangers are virtually impossible to repair, the furnace must be replaced.

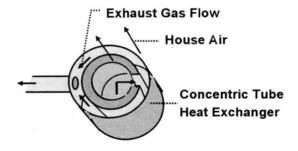

*Photo #19 shows a **concentric tube heat exchanger on an oil furnace** seen from the plenum. The inner and outer open rings is where warm air passes through to the house. The covered area — center circle and outer ring — is the top of the heat exchanger. The illustration shows a cutaway view and the flow of exhaust gases and air in the heat exchanger. The home inspector is not expected to open the plenum to view the heat exchanger. With this oil furnace, the heat exchanger was not visible from below. But we suspected a cracked heat exchanger due to the accumulation of soot at the room registers. We suggested that a furnace technician come in to examine it.*

#19 Concentric tube heat exchanger on an oil furnace

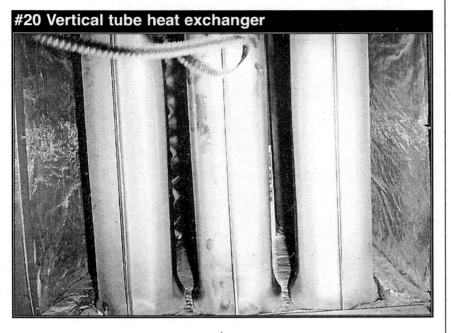

#20 Vertical tube heat exchanger

*Another type of heat exchanger, more commonly found in conventional gas furnaces, is the **vertical tube heat exchanger** as shown in **Photo #20.** This photo was taken after removing the register cover in the plenum (the strap is my camera strap). You can see welds at the top of each tube. The drawing shows a 4-tube heat exchanger. The wider chamber at the bottom of each tube sits over the burner and channels exhaust gases through the heat exchanger and out the smoke pipe. House air passes around and between the tubes, becoming heated, and passes into the house.*

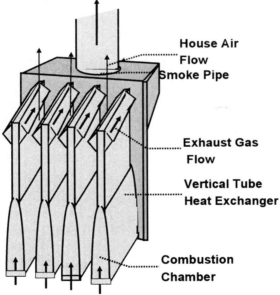

House Air Flow

Smoke Pipe

Exhaust Gas Flow

Vertical Tube Heat Exchanger

Combustion Chamber

- **Humidifiers:** The home inspector may find a humidifier added to forced warm air furnaces. The purpose of the humidifier is to add moisture to the circulating air when the furnace is operating. Humidifiers may be found mounted in the return duct or in the supply plenum. Those located over the heat exchanger can damage the heat exchanger if they leak, causing rusting and corrosion.

There are several types of humidifiers. The **drum humidifier** is located in the return duct with a bypass to the supply plenum. The **evaporative humidifier** can be found on the supply plenum with a tray of evaporative pads inside the plenum. This type is likely to develop leaks over the heat exchanger. Other kinds are cascade, atomizing, and steam generating.

The home inspector is not required to inspect the humidifier. However, you should check the humidifier for its general condition and leakage and for potential or actual damage to furnace components.

• **Supply and return ducts:** Heated air is blown through the furnace to room registers through supply ducts. Cool air is gathered in return grills for redelivery to the furnace. There are 2 basic types of distribution systems. The **extended plenum system** (shown below) consists of the plenum mounted on the furnace, a large extended supply duct, and branch ducts that deliver heat to registers in each room. Registers for outside rooms are normally located along a perimeter wall. Return grills, which should be located on walls opposite the heat registers, bring cool air back to the furnace through return ducts.

Another type of distribution system is called the **radial system**, where each supply duct takes off directly from the furnace to individual room registers. In this system, there is no main supply duct. A variation on the radial system is the **perimeter loop arrangement**, common in most slab houses, where one duct encircles the entire perimeter of the house and feeder ducts run from each register back to the furnace.

Supply and return ducts can be made of sheet metal, fiberglass, plastic, glass reinforced plastic sheets, or even wood. In any case, ducts should be air tight, properly

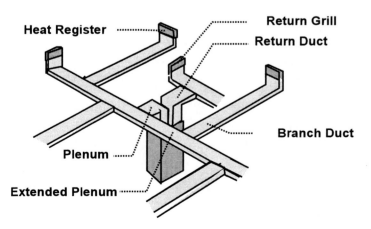

supported, and free of damage such as corrosion. Ducts should be insulated in cold areas, where they pass through an unheated attic or crawl space, for example.

For noise control and to reduce vibration, there may be a **canvas fabric collar** between the main supply duct and the furnace. The collar should be in good condition, not torn or having open sections. Some fabrics used in older systems contained asbestos, and the customer should be warned about that. Supply ducts may have **balancing duct dampers** used to equalize the volume of air in the branch ducts. If the distribution system is zoned, where supply and return ducts are divided into separate loops, the flow of heated air is controlled by **motorized zone dampers**. In this case, there would be an individual thermostat for each zone which causes its motorized damper to open and close.

The presence and placement of **return grills** is important for good air circulation. Furnace operation requires that air returning to the furnace through return ducts be about 80% of the supply air. Therefore, each area of the house that has a heat register should also have a return grill. For rooms without a return grill, doors should be undercut (about 1" of clearance above carpets) for good air flow. Ideally, there should be no **open return** in the basement since leaking combustion air from the furnace may be drawn into the home's air supply. Some local codes permit an open return in the basement if it's at least 10' away from the furnace. Check your local codes.

NOTE: Supply and return ducts in slab-on-grade homes are normally buried in or beneath the concrete slab, making inspection of them impossible for the home inspector. Ducts in the concrete can collapse or leak and rust out, causing a restriction in air flow. The inspector should pay attention to air flow at registers and grills to verify that air flow is not restricted. Customers should be informed that the inspector cannot inspect the ducts themselves.

- **Exhaust system:** As discussed in earlier pages, conventional forced air systems have a conventional smoke pipe exiting to the chimney. High-efficiency furnaces exhaust cooled gases in PVC piping through the house wall and drain condensate to the floor drain. All piping should be inspected for its condition.

CAUTION
There should be <u>no open return</u> in the basement. Leaking combustion air from the furnace could be drawn into the home's air supply.

Definitions

A <u>*balancing duct damper*</u>, *located in the branch supply ducts, equalizes the flow of warm air to the house.*

A <u>*motorized duct damper*</u>, *located in zoned supply ducts, controls the flow of warm air to zones within the house.*

For Your Information

Check your local codes on this issue of whether an open return is allowed in the basement within 10' of the furnace. It's not allowed in our area.

Inspecting Forced Warm Air Furnaces

The home inspector will inspect the warm air **furnace** itself, its **distribution system**, and its **venting system**. Note that the procedures for inspecting the fuel burner, smoke pipe, chimney, and cleanout were presented in earlier pages in this guide (see pages 24 to 31 for gas burners and pages 39 to 43 for oil burners). We won't repeat these procedures here. The home inspector should also inspect the forced warm air furnace for the following conditions:

- **Lack of heat source in each room:** As you inspect each room in the house, note whether a heat source is present and be sure to report if there is no heat source. After you've turned up the thermostat, you can hold a tissue in front of each heat register to check the flow of air from the register into the room. Lack of air flow can indicate a restriction in the supply ducts, a fan problem, or other furnace malfunction. A tissue held against the return air grill will test the pull of air back to the furnace — the tissue should be sucked against the grill.

- **Dirty, loose, or missing registers or grills:** As you inspect registers, note if there is soot and dark staining on the walls above them. This condition can indicate a cracked heat exchanger which is sending combustion byproducts into the home, which would be a **safety hazard**. If return grills are missing, check the door to see if there's adequate air flow under it.

- **Defects in the furnace jacket:** Examine the outside of the furnace jacket for damage, holes, corrosion, or rust. Scorch and burn marks, in the area of the heat exchanger can indicate a heat exchanger problem. Staining and soot around the burner would indicate spillage and downdrafts.

- **Cracked, rusted, or corroded heat exchanger:** After turning off the furnace, remove the access panels and examine as much of the heat exchanger as possible. (You may not have any access to the heat exchanger in an oil furnace or high-efficiency furnace). Use your flashlight and inspection mirror to inspect the heat exchanger for cracks. Be particularly careful about looking at welded areas and at curves. Some manufacturer's models have had a history of faulty heat exchangers and other problems including corrosion in some of the earlier high-efficiency models due to condensation (see charts on next page).

Personal Note

"I always stress making a note in the inspection report about the absence of a heat source in each room. It seems a small thing, but customers get angry when they discover a room without a heat register after moving in. If you miss this point, you'll only end up paying to have a heat source put in. I've paid for some myself."

Roy Newcomer

Lennox Pulse Furnace		
Brand: Lennox Pulse Furnaces 1981 through December 30, 1989 **Model**: G14 or GSR14 **Problem**: Heat exchangers are cracking. Have a Lennox technician examine. Lennox no longer manufactures Pulse Furnaces		
Vent pipes for natural gas or propane furnaces and boilers		
HTVP - High temperature Plastic Vent Pipes recall. Vent pipes are plastic; the vent pipes are colored gray or black; the vent pipes have names "PLEXVENT", "PLEXVENT II", OR "ULTRAVENT" stamped on the vent pipe or printed on stickers placed on pieces used to connect the vent pipes together. Call 800-758-3688.		
Manufacturer, Trade name	**Brand name**	**Model Numbers**
Burner Problems		
Heil-Quaker	Heil, Whirlpool, Tempstar, Dayton, Sears	NUGK, NULK, NUDK, NDLK, NRLF, NRGH, NRGF, NUGE, NDGE, NULE
Premature Cracked Heat Exchangers (These furnaces have exceeded their life expectancy.)		
Mueller	Mueller Climatrol	Prefixes 140-149, suffixes 75 or higher
Lennox	Lennox	G8, G9, G10, G11, G12
Safety Concerns		
Concern: Cabinet insulation deterioration. Contact local contractor or 717-771-6418	York	P2DP Serial #: EECM through EGEM

A cracked or faulty heat exchanger can be indicated by the following items and the home inspector should watch out for each one. Each situation should be reported as a **major defect**, and the home inspector should **recommend evaluation by a furnace technician before closing**. The customer should be informed that the only cure for a cracked heat exchanger may be replacing the furnace.

Furnace Upgrades
The following furnaces are on notice as needing upgrades.
Heil, Whirlpool, Tempstar, model NUGK, Serial numbers H540 and smaller
Armstrong Magic Chef, models EG6B and EG7B

*Photo #21 shows 3 **burn marks** on a gas furnace corresponding to the location of 3 tubes in the vertical tube heat exchanger. This is a clear indication that there are cracks in the exchanger at these tubes, and combustion gases are deflecting against the side of the furnace cabinet. When we felt this area, it was hot to the touch. Watch also for burn marks at the front of the unit around the burner area in a gas furnace and the back of the flame shield. Flames hitting these areas can indicate metal fatigue in the heat exchanger (as well as other causes such as a blocked flue).*

TELL YOUR CUSTOMER

Inform your customer that only <u>about 25%</u> of the heat exchanger in a conventional gas furnace is visible for inspection. In oil-fired and high efficiency units you may not be able to see the heat exchanger at all. The furnace should be evaluated by a qualified furnace technician.

---

*Photo #22 shows an inspector **using an inspection mirror** to examine the heat exchanger in a gas furnace. Here, you can see how the mirror and flashlight are used to see up into the combustion area to get as good a view as possible of the heat exchanger. In this photo, you can see the seams of the exchanger.*

— **Scorch and burn marks on the furnace jacket:**

#21 Burn marks

— **Staining around registers or on front of furnace:** Refer back to **Photo #11** which shows signs of puffback on an oil furnace. Staining here or on walls next to the room registers can indicate a problem heat exchanger.

— **Visible cracks:**

#22 Using an inspection mirror

— **Corrosion, rust, or scaling on or below the heat exchanger:**

#23 Scaling and buildup

Photo #23 shows **scaling and buildup** *of metal flakes below the draft diverter area, which is the large silver-colored box at the top. With this much flaking visible, you can assume the heat exchanger isn't in good condition. Refer back to* **Photo #5** *of the dismantled gas burner that shows serious corrosion due to condensation in the furnace. You can assume that the heat exchanger is in this same condition. Of course, you must always try to examine the heat exchanger. Don't just accept these signs of a problem.*

— **Unusual flame activity:** As described on earlier pages for gas and oil burners (pages 24, 25, 39, 40), unusual flame action can mean problems with the exchanger.

• **Faulty temperature differential:** You can perform a temperature rise test on the furnace. This is an optional test that should only be performed during the heating season. Use your thermometer to check the temperature of the supply air about a foot off the plenum and the temperature of the return air. The furnace plate will state the furnace's **temperature rise range**, generally a difference of 30° to 100° between warm supply air to cold return air. For example, supply air may be at 160° and return air at 70°, putting the difference within the range (160 – 70 = 90).

• **Dirty, damaged, noisy, or malfunctioning fan:** Examine the fan before you fire up the system, looking for signs of damage, missing blades, and dirt. After the furnace is fired up, listen to the fan as it kicks in. Noises and excessive vibrations can indicate loose bearings or mountings, a worn belt, or misalignment.

• **Dirty or missing air filter:** Dirty air filters can impede the flow of air into the furnace, carry dust into the heat exchanger, and allow the plenum to get hot enough to turn off the furnace in a pattern of short cycling. Always check

Personal Note

"I was inspecting a badly corroded Lennox G-12 furnace. The burners were so clogged up the flames were hardly coming out. I knew immediately that the furnace needed replacement. However, I continued in my inspection and inspected the entire furnace. Even when you know the furnace is a write-off, complete the inspection. That's what customers paid for, and they'll be upset if you stop right away."

Roy Newcomer

for the presence and condition of the filter. Conventional filters should be cleaned or replaced regularly.

CAUTION: If you find an **electronic air filter**, turn off the master furnace switch, which will usually turn off power to the filter too. Wait about 30 seconds to let the static charge dissipate before pulling the filter out for inspection. Otherwise, you're in for a shock, literally.

- **Presence of asbestos:** If an older furnace has a fabric collar between the furnace and the plenum, it might contain asbestos. The home inspector won't be able to say with any certainty if asbestos is present, but customers should be warned about the possibility. Insulated ducts should be examined for asbestos as well. Duct insulation containing asbestos doesn't necessarily have to be removed (see page 53).

- **Dirty or leaking humidifier:** The home inspector is not required to inspect the humidifier for operation, but you should check it out for its potential damaging effect on other furnace components.

*Photo #24 shows a **humidifier with mineral deposits**, hanging right below the cool air return. This unit has been leaking, and we found corrosion in the ducts and furnace because of it. The unit should be cleaned and repaired.*

#24 Humidifier with mineral deposits

#25 Humidifier in the cool air return

*Photo #25 shows a **humidifier in the cool air return**, but not above the furnace, which is preferable. Notice that this humidifier has a bypass to the supply duct. Incidentally, we felt that the clothes hanging here were too close to the furnace and advised the customer of that fact.*

- **Leaking, damaged, loose, or corroded ducts:** Examine visible supply and return ductwork for condition, noting any open joints, corrosion, and loose supports.

- **Open return in basement:** Report the presence of an open return in the basement, which is forbidden by modern building codes. CO from the furnace can be sucked into the home's air supply. This situation should be reported as a **safety hazard**.

- **Missing or malfunctioning dampers:** Malfunctioning balancing dampers can be discovered at the registers where air flow and warmth would be impeded. Motorized zone damper operation can be verified by operating each zone's thermostat (one at a time) and checking air flow. Note that zone dampers don't close completely, so there will be some air flow when the damper is closed.

A CAUTION: When you finish your furnace inspection, be sure to turn the furnace back on and return thermostat settings to their original settings. Leaving the furnace off in an empty house in winter can be a major mistake for the home inspector.

Reporting Your Findings

When you're inspecting the heating system, have your customer present. It's a smart practice. When the customer comes with you, you have an opportunity to fully explain the inspection and point out findings to the customer. Customer

Personal Note

"An inspector I know (not on my staff, fortunately) forgot to turn the furnace back on after an inspection. The house happened to be unoccupied. When the owners returned, they found the plumbing pipes frozen and burst. Of course, the inspector had to pay for the damage and repair — a very expensive mistake."
Roy Newcomer

knowledge is a big step toward the prevention of complaint calls later.

Keep a running dialogue going with the customer. Not everyone is familiar with furnace operation, and they may not understand what you're doing and the significance of your findings. Because furnace defects can be a serious safety hazard to the home's occupants, you want to be sure that the customer understands what you're saying. So **keep it simple**, but do talk about it. And pay attention to whether the customer understands.

During the inspection, be sure to explain the following:

- **What you're inspecting** — the burners, the heat exchanger, the fan, the ductwork, the smoke pipe, and so on.

- **What you're looking for** — rust, corrosion, gas leaks, downdrafts, dirty filters, cracks in the heat exchanger, and so on.

- **What you're doing** — testing air flow and return with a tissue, turning off the furnace to inspect the burners, testing the draft diverter with a match, inspecting the heat exchanger, and so on.

- **What you're finding** — signs of a lack of combustion air, cracks in the firebox, a leaking humidifier, a filled cleanout and blocked chimney flue, and so on.

- **Suggestions about dealing with the findings** — cleaning the burners, changing the air filter, replacing the entire furnace, having a furnace technician come in to evaluate the heat exchanger, having the oil burner serviced, and so on. But with this caution — don't make uneducated guesses about how repairs should be made.

Be sure to review the inspection report with your customer after the inspection. Even though you've been careful to communicate during the inspection, often times the customer will forget some or all of what you've said. Go through the inspection report page by page, pointing out where you've marked certain findings. This is especially important with technical systems such as heating, where the customer may not understand thoroughly. The review gives you another chance to test their understanding. Spend time pointing out where you've indicated major repairs, safety hazards, and items requiring replacement in the near future.

Filling in Your Report

Every home inspector needs an inspection report. **A written report** is the work product of the home inspection, and every home inspector is expected to deliver one to the customer after the inspection. Inspection reports vary a great deal in the industry with each home inspection company developing its own version. Some are considered to be excellent, while others are not very good at all. A workable and easy to use inspection report is important for a home inspector in terms of being able to fill it in. Of greater importance are its thoroughness, accuracy, and helpfulness to your customer. We can't tell you what type of inspection report to use, but let's hope that it's professional.

The **Don't Ever Miss** list is a reminder of those specific findings you should be sure to include in your inspection report. We list these items after years of experience performing home inspections. Missing them can result in complaint calls and lawsuits later. Here is an overview of what to report on during the inspection of the furnace:

- **Furnace information:** Record the brand name, model, and serial number of the furnace. This should be located on the data plate on the furnace, although you may not be able to find all this information on an old gravity warm air furnace. Identify the type of furnace (gravity or forced) and the type of fuel used.

- **Operation:** Note in your report whether the furnace fired or not and make a special note if the furnace was not operating. If you performed a temperature rise test, you might want to report if the differential was within range.

- **Heat exchanger:** Note whether you were able to inspect the heat exchanger or not. Report any defects found such as rusting and corrosion, visible cracks. It's a good idea to give the heat exchanger an overall rating. Here's a suggestion on how to do that:

— Use **satisfactory** only if you've found no defects in the heat exchanger or any signs of defects *and* the furnace is not past its lifetime *and* it's not in the problem charts on page 69.

— Use **marginal** if you've found some signs that *might* point to a cracked heat exchanger *or* it's on the problem charts

---

DON'T EVER MISS

- Heat source in each room
- Burner problems
- Evidence of CO production
- Gas leaks, oil odors and smoke
- Holes, cracks, or rusting in heat exchanger
- Missing filters
- Upgrade or problem furnaces (charts)
- Improper clearances
- Spillage from draft hood or diverter or damper
- Improper venting and holes in vent pipes

Report Available

The American Home Inspectors Training Institute offers both manual and computerized reports. These reports include an inspection agreement, complete reporting pages, and helpful customer information. If you're interested in purchasing the <u>Home Inspection Report</u>, please contact us at 1-800-441-9411

or past its lifetime (even if you haven't found any problems). Recommend that the a technician evaluate the furnace before settlement if you rate it as marginal.

— Use **poor** if you've found cracks or holes in the heat exchanger *or* if you've found any of the other signs that indicate a cracked heat exchanger. Again, recommend that a technician evaluate the furnace before settlement if you rate it as poor.

- **Other components:** Report on the condition of the furnace jacket, burner, filter, fan, vent pipes, and distribution ducts and registers. Note defects such as oil burners that need servicing, fan motors with loose bearings, dirty filters, and open joints in vent pipes, for example.

- **Recommending an evaluation:** The home inspector should always recommend that a qualified technician be called in to evaluate the furnace if the following conditions are found. Be sure to make a note of your recommendation in the inspection report.

— If you rate the condition of the heat exchanger as marginal or poor.
— If the furnace didn't fire.

- **Major defect or repair:** If the furnace needs replacement either because of a cracked heat exchanger or if it's not operating, identify the condition as a major defect or repair in your report. Another major repair to list is the lack of return ducts in a gravity furnace.

- **Safety hazards:** Never miss reporting any safety hazards you've found. It's a good idea to report them on the furnace page of your report and then summarize them on a summary page at the back of your report. For the inspection of the furnace, don't miss these safety hazards:

— Cracked, rusted, or deteriorating heat exchanger
— Evidence of CO production
— Spillage, improper venting, holes in vent pipes
— Improper clearances
— Gas leaks and improper LP gas tank location
— An open return in the basement
— Blocked flue

WORKSHEET

Test yourself on the following questions.
Answers appear on page 78.

1. What component can be found on a forced warm air furnace but <u>not</u> on a gravity unit?

    A. A heat exchanger
    B. A bonnet
    C. A blower unit
    D. A burner

2. What would <u>not</u> be considered a major defect in an octopus furnace?

    A. Lack of return ducts
    B. Fuel inefficiency
    C. A cracked heat exchanger

3. Write the number of the item in the drawing below for each component listed.

    A. Burner area   _____
    B. Fan   _____
    C. Supply duct   _____
    D. Heat exchanger _____
    E. Fan motor   _____
    F. Air filter   _____
    G. Smoke pipe   _____
    H. Return duct   _____

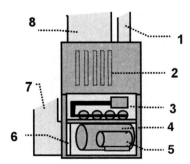

4. What is a plenum?

    A. A large supply duct over a furnace
    B. An auxiliary electric heater on oil furnaces
    C. A fan that pulls combustion products out of the furnace
    D. A component found only in pulse furnaces

5. In which type of furnace would you find a tailpipe?

    A. A condensing furnace
    B. A conventional furnace
    C. A gravity warm air furnace
    D. A pulse furnace

6. What is the purpose of the high limit control on a forced warm air furnace?

    A. It heats the air.
    B. It turns off the burners if the furnace hits a preset high temperature.
    C. It turns on the fan when the furnace hits a preset high temperature.
    D. It turns off the fan when the furnace reaches a preset low temperature.

7. What should the home inspector do if he or she finds a cracked heat exchanger.

    A. Report it as a major repair.
    B. Sell the customer a new furnace.
    C. Turn off the furnace immediately.
    D. Tell the customer it will probably last 5 more years.

8. What is <u>not</u> caused by a dirty air filter?

    A. Impeded air flow
    B. A dirty heat exchanger
    C. Short cycling
    D. Excessive vibrations

9. What finding would <u>not</u> indicate a faulty heat exchanger?

    A. Burn marks on the furnace jacket in the area of the heat exchanger
    B. A leaking humidifier over the heat exchanger
    C. A corroded smoke pipe
    D. Metal flakes in the burner area

Guide Note

*Pages 78 to 95 present the study
and inspection of hot water
heating systems.*

Chapter Seven

HOT WATER BOILERS

Although the term *boiler* is used to describe the heating unit, the water in the hot water heating system boiler is not actually brought to the boiling point. The water in a hot water system is heated to about 160°F to 185°F (water boils at 212°F).

Types of Hot Water Systems

A hot water heating system heats a home by convection. That is, water is heated in a boiler and moves through pipes to radiators or convectors throughout the house, transferring heat with it. The water then cools and returns to the boiler to be reheated and re-circulated. This pattern of water circulation and re-circulation is the basic principle of hot water systems. A hot water system is completely filled with water.

- **Gravity hot water systems:** The earliest hot water systems, which are no longer installed, are called gravity systems. When heated, water expands and becomes lighter so that it rises. Since the system is filled with water, the rising hot water pushes the cooler water in the piping ahead of it. This starts a simple cycle of circulation through a gravity system.

  Piping in the old gravity hot water systems is about 3" in diameter; modern hot water systems use piping that is smaller — about 1" in diameter. A gravity system has no moving parts.

  Water expands as it's heated, and a gravity system has an **expansion tank** to accommodate this expansion. Most often, the expansion tank in a gravity system is open to the atmosphere. That means that the tank will overflow to an **overflow pipe**, routed to the outside or to a basement drain, when there's too much water in the system. This is called an **open system.** Water in an open system is not under pressure. The expansion tank is located above the highest radiator in the home, typically in the attic, and has a sight glass to show water level. When water in the gravity hot water system is low, water is added manually through the **manual fill valve** (when the boiler is cold).

Worksheet Answers *(page 77)*

1.   *C*
2.   *B*
3.   *A is 3.*
        B is 4.
        C is 8.
        D is 2.
        E is 5.
        F is 6.
        G is 1.
        H is 7.
4.   *A*
5.   *D*
6.   *B*
7.   *A*
8.   *D*
9.   *C*

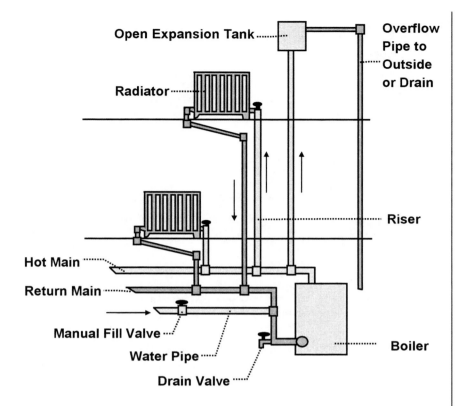

Open Expansion Tank

Overflow Pipe to Outside or Drain

Radiator

Riser

Hot Main

Return Main

Manual Fill Valve

Water Pipe

Drain Valve

Boiler

Definitions

In a *hot water heating system*, water heated in a boiler is transmitted through pipes to radiators. Heated water rises naturally in a *gravity hot water system*. Water is circulated in a *forced hot water system* by a circulating pump.

An *expansion tank* in a hot water system provides space for water to expand into. An *open system* has an expansion tank open to the atmosphere and located above the highest radiator. A *closed system* has a sealed expansion tank located just above the boiler.

Gravity hot water heating systems are obsolete but may still be found in operation in older homes. The old boiler is typically made of cast iron and is 2 to 6 times larger than the new modern boilers. Their heating efficiency is low. Old cast iron radiators are usually found with these systems. A radiator control valve at one end of the radiator regulates water flow, and therefore heat, to the radiator.

- **Forced hot water systems:** A forced hot water heating system is also called a **hydronic** system. It can be recognized by the presence of a **circulating pump**. In this system, water circulation doesn't depend on natural processes as in a gravity system. Here, water is moved through the piping system by the action of the pump.

A forced hot water system is a **closed system**, meaning that it has a **closed expansion tank**. This tank is partially filled with trapped air which compresses to accommodate the expansion of heated water. The tank is located near and just above the level of the boiler.

In a closed system, the water everywhere in the system is under pressure. The pressure is about **12 psi** (pounds per square inch) in the typical 2-story home. This is based on the principle that 1 psi of pressure is required to lift water

2.31' of height. Therefore, it takes over 8 psi to lift water the required 19' in a 2-story home. Add a few extra pounds to maintain pressure in the upper-floor radiators, and the total pressure required would be about 12 psi. (3-story homes would require 17 psi, 4 stories 22 psi, and 5 stories 26 psi.)

This internal system pressure is maintained by an **automatic fill valve** that adds water to the system to keep it at the required pressure and a **pressure relief valve** that discharges water from the system when pressure reaches a dangerous level. A **pressure reducing valve** reduces the higher pressure of water in the plumbing system before it reaches the boiler.

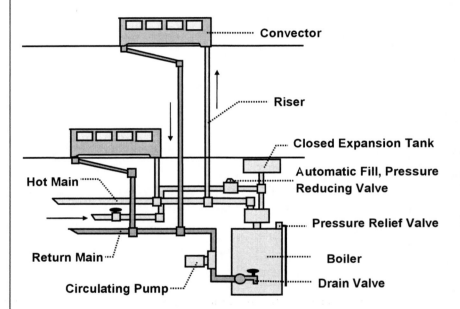

The newer forced hot water systems use baseboard or freestanding **convectors** instead of the old radiators. Convectors have a water tube running through them. Attached to the tube are a number of fins or metal plates that heat the air passing through the convector.

- **2-pipe systems:** A forced hot water heating system may have separate piping runs to and from the boiler. Heated water is delivered to convectors through the hot main and risers; cooler water is returned to the boiler through the return piping (as illustrated above). Although this system is more costly to install, it allows for **zoned heating**. Zoned heating is accomplished with the addition of zone valves or multiple circulating pumps or even with multiple boilers in larger homes.

- **1-pipe systems:** Forced hot water systems may only have a single piping run. The **series loop** (shown at the top right) is one continuous piping run that incorporates the radiators or baseboard convectors into the run. Closing off one convector in the run would shut off heat supply to the whole loop. Another approach with a 1-pipe system is to have the convectors attached to the run with risers, allowing each one to be controlled without affecting water flow in the loop (shown in the second illustration.) With either approach, the convectors at the end of the run receive cooler water than those at the beginning. To compensate for this, convectors are usually smaller near the boiler and larger at the end of the run.

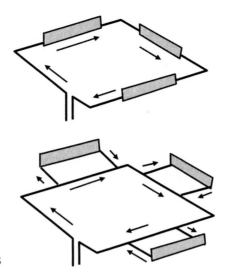

- **Radiant panel heating:** In some hot water systems, water is circulated through continuous rows of piping called panels, which are buried in the floor or ceiling. The piping heats the floor or ceiling which radiates the heat to the room. Older radiant panels used black steel or galvanized steel pipes. Today, pipes are more likely to be copper or plastic tubing. This system can be zoned to supply heat to different sections of the home. The home inspector may find radiant hot water systems in slab-on-grade homes, where the piping is buried in the concrete slab.

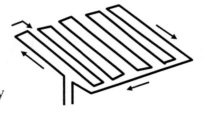

For human comfort, water temperatures in radiant hot water heating are kept lower than in other types of hot water heating systems. The temperature of the water ranges from about 85° for floor installations to between 115° and 120° for ceiling installations.

Definitions

A *1-pipe hot water system* uses a single piping run for moving water to and from the boiler. A *2-pipe system* has separate supply and return piping runs.

Radiant panel heating is a hot water system consisting of continuous piping laid out in rows which are buried in the floor or ceiling.

Types of Hydronic Boilers

Just like forced air furnaces, there are generations of hot water boilers from the older units to the newest high-tech units. The conventional forced hot water, or hydronic, boiler has these basic components:

- The **burner area** and combustion chamber, located at the bottom of the boiler.

- The **heat exchanger** above the burner from which exhaust gases are vented and the surrounding water is heated and then distributed to the home.

- The **circulating pump**, which sits outside the boiler on the return pipe, is used to move water through the heating system.

- On the outside of the boiler is a **pressure relief valve** that provides an escape for hot water if pressure should increase beyond allowable limits. There is also a **temperature, pressure gauge** showing readings of boiler conditions.

- A coil, called the **tankless coil**, may be inserted in the boiler to heat water for domestic use.

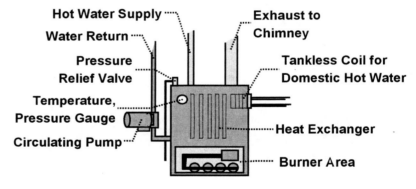

Conventional Boiler

Removing the access panel on the burner side of the boiler will give the home inspector access to the burner area and the heat exchanger casing. But in a boiler, the heat exchanger is not visible and cannot be inspected directly.

Some forced hot water boilers are powered by electricity. The **electric boiler** operates similar to the electric water heater. Electric resistance coils, insulated against the water, are inserted through the boiler wall and made leak tight. The heating elements are energized in sequence to prevent an electrical overload. When the thermostat is satisfied, the elements and the circulating pump are turned off. Smaller electric boilers may be used to heat add-on rooms.

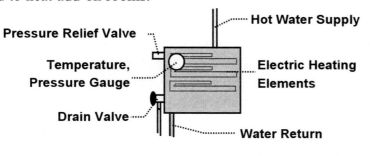

Electric Boiler

The **mid-efficiency boilers** that came on the market in the mid-1970's marked improvements in much the same way as furnaces were improved at that time. They're basically conventional boilers with the addition of **an induced draft fan** to pull combustion gases through the unit and thus reducing heat loss. **Motorized vent dampers** were also added to prevent warm air from escaping up the chimney when the boiler was off.

High-efficiency boilers can perform at 95% efficiency. Fired by either gas or oil, they vary in configuration from manufacturer to manufacturer. **Condensing boilers** have 2 or more heat exchangers to extract as much heat as possible from the exhaust gases, condensing a great portion of the gases. The condensate is drained to the floor drain, while the remaining cool gases are forced with an induced draft fan through 3" PVC piping to the outside through the house wall.

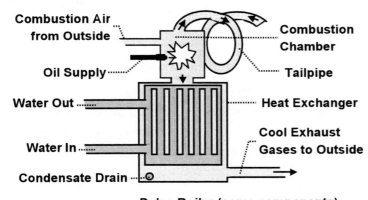

Pulse Boiler (some components)

The high-efficiency **pulse boiler** (shown above) can be gas or oil-fired. The principle of combustion is the same as that of the pulse furnace (see page 60 for an explanation). As with condensing boilers, condensate from the pulse boiler is drained to the floor and the remaining cooled exhaust gases are vented to the outside through PVC piping.

- Thermostat
- Pump control
- Limit control
- Pressure relief valve
- Pressure reducing valve
- Automatic fill valve
- Low water cut-off

Definitions

The pump control turns on the circulating pump at a signal from the thermostat upon a call for heat or from an aquastat at a certain preset temperature.

The limit control turns off the burners if the water reaches a preset high temperature.

An aquastat is a temperature-sensitive device, immersed in water to detect water temperature, that activates the circulating pump. A modulating aquastat senses outdoor temperature and changes circulating water temperature requirements accordingly.

Boiler Components

Forced hot water, or hydronic, boilers have the following components:

- **Boiler controls:** The **thermostat** is one of the main operating controls of a hot water heating system. In addition to the thermostat, there are a number of operating and safety controls.

In newer boilers, because the heat exchanger is smaller and more efficient, the thermostat will activate the burner and circulating pump simultaneously. In some boilers, however, the pump can be manually set to run constantly during the heating season. In older systems, when there is a call for heat, the thermostat activates the burner only. A separate **pump control** with a temperature-sensitive **aquastat** turns on the pump when the water temperature reaches a preset level of about 120°. Some systems have a **modulating aquastat** that senses outdoor temperatures and turns on the pump at varying water temperatures. The lower the outside temperature, the higher the circulating water temperature.

If the hot water boiler is also used as a source of domestic hot water, then the thermostat will control only the circulating pump. The burner is controlled by an aquastat that senses water temperature. In such systems, a flow-control valve is present that prevents hot water from rising into the distribution piping when heat is not called for in the house.

A hot water boiler has a safety device called the **limit control**, which will turn off the burner if temperatures get too high. If water was allowed to boil in a hot water system, expanding steam could burst pipes and damage the boiler. Therefore, the limit control is set at about 200°, below the boiling point of water.

Another safety control on the boiler is the **pressure relief valve**. This valve prevents pressure within the system from exceeding a dangerous level (usually set at 28 psi to 30 psi). Because the water discharged from a pressure relief valve is very hot, an extension should be attached to it to discharge within 6" of the floor. This extension should not be threaded at the bottom or be capped.

Another safety control found on some larger boilers is the **low water cut-off**. This device will shut down the burners if there isn't enough water in the system.

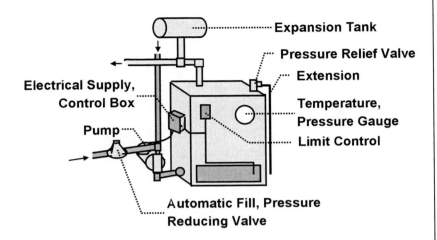

Definitions

A <u>pressure relief valve</u> will discharge water from a boiler if pressure approaches dangerous limits.

The <u>pressure reducing valve</u> reduces water pressure in the plumbing line to acceptable boiler pressure. An <u>automatic fill valve</u> adds water to the boiler if pressures fall below the required pressure.

A <u>low water cut-off</u> shuts off the boiler when water levels fall.

There are other controls on the plumbing line to the boiler. A **pressure reducing valve** will automatically reduce the pressure on incoming water to the right level for the heating system —12 psi for a typical home. The **automatic fill valve**, part of the same unit as the pressure reducing valve, adds water to the system if pressure falls below 12 psi. In the newest systems, this unit will also be equipped with a **back-flow preventer** that prevents water from the heating system from backing up into the plumbing system.

All hot water boilers have a **temperature/pressure gauge**, either in a combination gauge or as a pressure gauge plus a pencil-type thermometer. The pressure showing is the actual working pressure of the system. If this pressure exceeds 30 psi and the relief valve is not discharging, the boiler should be turned off.

- **Zone controls:** In zoned heating, where supply piping is divided into separate loops, each zone has its own thermostat. The thermostat first turns on the burner on a call for heat. Then it activates either the zone's circulating pump (if each zone has one) or the zone's **zone valve** (if there's only one circulating pump). The zone valves are electrically operated.

For Beginning Inspectors

It's time to see some boilers. For older systems, try any friends that live in older homes. You may even find an old gravity system. For newer systems, try a heating outlet and get someone to show you the new models. Since there's variation from model to model, try to locate and identify boiler controls.

- **The circulating pump:** On a hydronic or forced hot water boiler the circulating pump, located in the return piping, moves the water through the boiler and piping system. Depending on the type of boiler, the pump is designed to operate constantly, intermittently only when heat is called for, or at certain preset water temperatures. Constant operation can be controlled by a manual on/off switch, set to "on" during the heating season and to "off" during the summer.

Take a look again at **Photo #9**, showing the converted oil-fired hydronic boiler. Note the circulating pump at the side of the boiler. This is a standard Bell & Gosset pump. It should be oiled occasionally during the heating season, but not over-oiled, and should be running smoothly without any leaks. Notice the safety or serviceman's switch at the upper left of the boiler side. In this unit, the thermostat connection and controls to the burner and pump are in this same control box.

#26 Zoned heating system

*Photo #26 shows a **zoned heating system** for a large home. Here, there are 3 hydronic boilers, each with its own circulating pump. The pumps can be seen at the left of each boiler. Notice that all 3 boilers are vented to the same flue.*

#27 Unique gas boiler

*Photo #27 shows a **unique gas boiler** hanging on the chimney. The circulating pump is at the left, and you can see the pressure reducing valve in the fill pipe just after a manual shut-off valve (upper left). Notice the large draft hood on the boiler. You can also see evidence of corrosion on the smoke pipe in this photo.*

- **Burner area and combustion chamber:** See pages 24 to 31 of this guide for information on gas burners and pages 39 to 43 for oil burners. Electric boilers, of course, do not have a burner or combustion chamber area. Instead, you'll find a bank of heating elements as shown on page 83.

- **Heat exchanger:** The boiler heat exchanger is not visible for inspection. The heat exchanger can be made of cast iron, copper, or steel. It allows the heat from the burner to pass through it while it heats the water on the other side of the heat exchanger. A boiler may be a **dry base boiler**, where the heat exchanger is set above the combustion chamber, normally seen in gas-fired boilers with vertical tube cast iron heat exchangers.

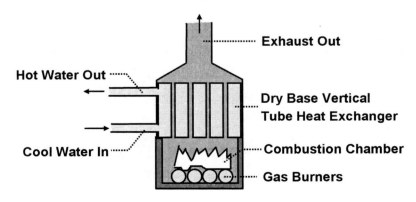

A **wet base boiler** is constructed so that water in the heat exchanger surrounds the combustion chamber. A wet base horizontal tube steel heat exchanger is primarily used with oil-fired boilers.

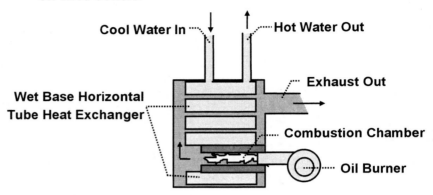

The same danger that exists with forced warm air heat exchangers, where combustion byproducts can enter the home's air supply, does not exist with boiler heat exchangers. If a heat exchanger develops holes from rusting or corrosion, water will leak out into the combustion chamber. The home inspector can spot a cracked or rusted heat exchanger by the presence of water damage in the burner area. A heat exchanger may be repaired in rare cases, but most often the boiler must be replaced.

- **Expansion Tank:** An expansion tank provides space for heated water to expand into. As mentioned earlier, the **open expansion tank** for a gravity hot water system will be located above the highest radiator, most likely in the attic. These tanks have an overflow pipe that will discharge excess water to the outside, sometimes onto the roof, or to a floor drain in the basement.

A forced hot water system has a **closed expansion tank** that is partially filled with air. When water expands in the closed system, the air in the tank is compressed. The tank is located above the level of the boiler so that air cannot get into the supply piping. (Any air trapped in the circulating water is removed by a trap at the top of the boiler that should be bled off manually from time to time.)

One type of expansion tank has water and air in contact with each other. As the water absorbs the air over time, the tank can become **waterlogged**. Newer tanks have a diaphragm between water and air to prevent air absorption, although there can be problems with the diaphragm to cause a waterlogged condition. A waterlogged expansion tank causes the pressure relief valve on the boiler to discharge whenever the boiler is on due to increased pressure. The tank should be flushed and air put back in.

- **Distribution piping:** Supply and return lines for a forced hot water heating system should be well supported and free of leaks. Pipes passing through unheated areas such as crawl spaces and attics should be insulated. See pages 80 and 81 for information on the various distribution layouts such as 1-pipe and 2-pipe systems and zoned systems.

  With **radiant panel heating**, the pipes can leak into the floor or ceiling. Such leaks are difficult to locate and repair. Repairs require the floor or ceiling to be opened and can be very costly. A break in the piping can leak for a long time before it's noticed. One sign of a problem can be that the automatic fill valve is sending a constant stream of water to the system. The valve will be cold to the touch, and there will be an audible noise at the valve.

- **Radiators and convectors:** Cast iron radiators allow water to pass through, heating the metal and radiating heat in all directions. A control valve at one end of the radiator can be shut off to prohibit water flow through the radiator. Old valves may not shut off completely or will leak if adjusted. A small bleed valve is located near the top of the radiator allowing trapped air to be removed periodically,

  Convectors have a water tube through them. Attached to the tube are a number of fins that heat up and heat the air passing through the convector. **Baseboard convectors** are normally copper tubes and aluminum fins. A sheet metal housing with openings and vanes above and below the pipe help increase air flow. The baseboard convector should be installed high enough above the carpeting to allow air flow. Some control of heat to the room is normally achieved by closing a vane to restrict air flow through the convector.

- **Exhaust system:** As discussed, gas and oil-fired heating units must have an exhaust system to rid the unit of combustion byproducts. Old gravity and conventional forced hot water boilers exhaust through the chimney, while the high-efficiency boilers exhaust cooled combustion gases in PVC piping through the house wall and drain condensate to the floor drain. All exhaust piping should be inspected for condition and proper venting.

Personal Note

"I do encourage you to develop a library of home inspection books. With boilers and furnaces, there are so many models available that you can't expect to see them all before you start inspecting. The book by James Kittle, mentioned on page 1 of this Guide, is really helpful. It's got good photographs and detailed explanations of boilers and boiler parts."

Roy Newcomer

Inspecting Hot Water Boilers

When inspecting the gravity and forced hot water boiler, the home inspector will be inspecting:

- The **boiler** itself including the fuel burner
- The **distribution piping** and registers or convectors
- The **venting system** including the smoke pipe, chimney, and cleanout

The procedures for inspecting the fuel burner, smoke pipe, chimney, and cleanout were presented in earlier pages in this guide (see pages 24 to 31 for gas burners and pages 39 to 43 for oil burners). We won't repeat those procedures here.

The home inspector should also inspect the hot water boiler for the following conditions:

- **Lack of heat source in each room:** Again, as you inspect the home interior, turn up the thermostat, check for a heat source in each room and that it is functioning. Radiators should be located on exterior walls, preferably under windows. Sometimes, these old radiators may not be working or are disconnected. Be sure to feel each for the presence of heat. For zoned heating, be sure to turn up each thermostat and verify heat in each zone.

- **Leaking radiators and convectors:** As you check the heat source in each room, examine each radiator to see if its control valve is leaking but **do not operate this valve**. Note if the air bleed valve has been painted shut, which often happens. Point out that the valve should be kept in operating condition. Examine convectors for general condition too, reporting if their fins are dirty, if carpeting restricts air flow through the convector, and if any leaking is present.

- **Leaking radiant panels:** If you determine that the home is heated with hot water radiant panels in the floor or ceiling, keep an eye out for any leaking as you move through the living area.

- **Defects in the boiler jacket:** Inspect the outside of the boiler jacket for leaking, damage, holes, corrosion, or rust. Carefully check the boiler for leaks and dripping.

Excessive rusting or staining in an area can also indicate a leak in the inside. A leaking boiler may be able to be repaired, but it will probably have to be replaced. Leaking should be reported as a **major defect**. Scorching or burning can indicate a problem with the combustion chamber. Staining and soot around the burner would indicate spillage and downdrafts.

- **Temperature or pressure too high or low:** Read the temperature, pressure gauge while the boiler is still operating. Note if the temperature is too high for safe operation (over 200°F) and if the pressure in the boiler is too high (over 30 psi) or too low (under 12 psi for a typical home). If the temperature or pressure readings are over these indicated levels, *turn off the boiler immediately* and report the condition as a **safety hazard**. Be sure to inform the homeowner to have the boiler inspected by a qualified technician before re-firing the unit. The cause of the problem can be a faulty pressure relief valve that should be discharging water. Pressures above normal may signal a waterlogged expansion tank. During your reading, note if the gauge itself is broken, cracked, and so on.

- **Leaking, corroded, or missing pressure relief valve:** Inspect the relief valve, which should be located on top of the boiler or on the side near the top, not along the piping away from the boiler. ***Do not operate this valve.*** The valve should be free of corrosion and should not leak. Missing pressure relief valves should be reported as a **safety hazard**. Leaking or corroded valves should be replaced.

Check that the extension is present and properly installed. The extension pipe diameter should not downsize as it goes down the boiler. Check that the extension ends about 6" from the floor and is not threaded or capped at the end. Improper extensions should be reported as a **safety hazard**. If there's evidence that water has been discharging from the relief valve frequently, it may be an indication of a waterlogged expansion tank. High pressures would build up each time the boiler is fired, causing the relief valve to discharge water.

CAUTION
If the temperature, pressure gauge shows readings <u>above 30 psi or 200°</u> for a forced hot water boiler in a typical home, turn off the boiler immediately.

• **Cracked or rusted heat exchanger:** Use the safety or serviceman's switch to turn off the boiler. Then remove the access panel on the burner side of the boiler. The heat exchanger will not be visible for inspection, but as you examine the burner area, note if there is any evidence of leaking from the heat exchanger onto this area. There would be signs of rusting, corrosion, flaking metal, or water seepage and even dripping water. This condition should be reported as a **major defect** in the boiler. Suggest that a qualified service technician examine the boiler and point out to customers that the boiler will probably have to be replaced.

• **Dripping or malfunctioning circulating pump:** After checking the interior of the boiler, fire it back up to proceed with your inspection of the burners. At this time you'll be able to determine if the thermostat turns on the pump and burners simultaneously. The pump should not be leaking or dripping. Listen to the pump for noises that can indicate worn bearings. Note if the pump appears to be over or under oiled. Circulating pumps only last about 10 to 15 years and have to replaced eventually. Report if the pump is not operating at all.

• **Buzzing or arcing electrical work:** As the boiler operates, listen to the electrical wiring for any signs of buzzing or arcing. Report if any of the control boxes on the boilers are missing their covers — they should be covered.

• **Leaking or loose tankless coil:** If the boiler has a tankless coil insertion to produce domestic hot water, examine it to see if it is leaking or if nuts or bolts connecting it to the boiler are loose. Another type of domestic hot watersystem you may see is the **sidearm heater** which is a tankless coil in a separate gravity fed piping loop at one side of and above the boiler. The tankless coil should have its own pressure relief valve.

• **Waterlogged or leaking expansion tank:** As the boiler re-fires and continues to operate, note if pressure is quickly building up to the point where the relief valve discharges. This is an indication that the expansion tank is waterlogged. Even diaphragm tanks can become waterlogged. Inform the customer of this condition and recommend that the expansion tank needs to be checked and have air put back in. Check the exterior of the tank for leaks as well.

- **Leaking zone valves:** If the heating system is zoned and has zone valves rather than separate circulating pumps, check the zone valves for leaks and drips.

- **Leaking or poorly supported piping:** Examine visible piping for leaks. Make a note of loose, broken, corroded, or missing supports. Check to see if piping is insulated in unheated areas, and if the insulation contains asbestos, point that out to the customer. Suggest that asbestos be removed if in poor condition.

- **Noise in the heating system:** Lastly, listen for noises. Water circulation in hydronic systems should be noiseless. Noisy water circulation indicates air in the pipes that should be bled off. Squeaks in the pipes just after water begins to circulate indicates thermal expansion of the pipes and rubbing against supports. Support hangers can be oiled or lined to eliminate this problem.

A CAUTION: When you finish your boiler inspection, be sure to turn the boiler back on and return thermostat settings to their original settings. Leaving an empty house without heat in the winter can be a major mistake for the home inspector.

Reporting Your Findings

Talk to your customer while you're inspecting the boiler, but have them stand back as you fire it. Always consider yourself responsible for your customer's safety.

Be sure to explain to your customer what you were inspecting at the boiler and what you found. Take the time to answer questions. Remember that customers may not understand what they see at the boiler and are counting on you to make sense of it for them. Be sure to stress safety hazards you find and tell the customers you will indicate safety hazards in the inspection report for them. Suggest that customers review the report again on their own.

WHAT'S THE SYSTEM?

If a hot water boiler has no circulating pump, you've found an old gravity hot water heating system.

For Beginning Inspectors

Get out there and inspect some hydronic boilers. Follow the procedures as instructed in these pages, being careful not to operate those valves we've said not to operate. Don't forget to inspect the burner, the vent piping, chimney, and cleanout as well (see earlier pages on gas and oil burners for procedures). If your friends are willing, let them pretend to be your customers so you can practice your communication skills.

Note these items as items
requiring replacement
within the next 5 years in
your inspection report:
They will be reaching the
end of their lifetime in 5
years.

- Cast iron and steel
  boilers over 30 years old

- Copper boilers over 10
  years old

- Circulating pumps over
  10 years old

When reporting on the inspection of the hot water boiler, be sure to report on the following for both gravity and forced hot water boiler, gas and oil-fired:

- **Boiler information:** Record the brand name, model, and serial number of the boiler. This should be located on the data plate on the boiler, although it may be missing from an old gravity system. Identify the type of boiler (gravity or forced hot water) and the type of fuel used.

- **Operation:** Note in your report if the boiler fired or not and make a special note if the boiler wasn't operating.

- **Heat exchanger:** Note that the heat exchanger is not visible for inspection. But report any evidence of defects found such as leaking and rusting and corrosion in the burner area. Give the heat exchanger an overall rating since the boiler will most likely have to be replaced if heat exchanger problems exist. Here's a rating system:

— Use **satisfactory** if you've found no signs of leaking from the heat exchanger into the burner area *and* the boiler is not past its lifetime.

— Use **marginal** if you've found some minor signs of a problem *or* the boiler is past its lifetime. Recommend that a qualified technician come in to evaluate the boiler if you find the boiler marginal.

— Use **poor** if you've found definite and extensive signs of a defective or leaking heat exchanger. Recommend a technician if you rated the boiler as poor.

- **Other components:** Record your findings on the condition of the boiler jacket, burner, circulating pump, gauges and valves, piping and expansion tank, radiators or convectors, and venting systems. Write up the defects you've found such as a leaking pressure relief valve, a waterlogged expansion tank, a pump that needs oil, and oil burners that need adjustment, for example.

- **Major defects and repairs:** Boiler conditions that should be especially noted for major defect or repair include a faulty heat exchanger, a seriously leaking boiler, a boiler operating with temperatures or pressures too high, or a boiler that is not operating. These conditions should be noted in your inspection report.

- **Safety hazards:** All safety hazards found during the inspection of a hot water boiler should be noted in the inspection report.

— Gas leaks
— Improper clearances
— Spillage, improper venting, holes in vent pipes
— Boiler temperature or pressure readings too high
— Missing pressure relief valve or improper extension

NOTE: We stress the importance of accurate and detailed reporting because of the high-ticket liability the home inspector has regarding the heating system. We're sure you get the idea. If you miss reporting findings listed in the Don't Ever Miss list, you know you'll hear about it later in the form of a complaint call. The home inspector who misses these details in the inspection report will only have to pay later — perhaps for a whole new boiler.

DON'T EVER MISS

- Heat source in each room
- Burner problems
- Gas leaks, oil odors and smoke
- Leaking boilers
- Dangerous temperature or pressure readings
- Missing relief valve or improper extension
- Defective and leaking heat exchanger
- Improper clearances
- Spillage from draft hood or diverter or damper
- Improper venting and holes in vent pipes

Personal Note

"No, I don't have any stories about boilers blowing up. My inspectors know to turn off the boiler if gauge readings are too high and the pressure relief valve is not discharging. They've learned to be careful, as I hope you will."

Roy Newcomer

Chapter Eight

STEAM BOILERS

A steam boiler is similar to a hot water boiler in many ways, although there are some basic and important differences.

Steam Systems

When water is heated to the boiling point (212°), it changes to steam which contains heat. Steam rises through a large diameter piping system to radiators in the home, pushing air ahead of it. When heat is given up at the radiator, the steam condenses back into water which returns through piping to the boiler. Air valves on the radiators and/or vents on the return piping allow air to escape. This is a **gravity system**. Steam systems are no longer installed as a new installation, although the home inspector may see replacement boilers and new parts installed to keep an old steam system in operation.

Steam boilers operate with the boiler **3/4 full of water**. The rest of the boiler and the piping system are filled with air when the boiler is at rest. Pressure within the boiler is much lower than that of hot water systems, only from about **0.5 psi to 3 psi** as opposed to 12 psi in hot water boilers.

- **1-pipe system:** A 1-pipe steam system (shown below) has a single pipe attached to each radiator. Steam flows into the radiator through this pipe, and condensate returns down the same pipe. Radiators have a **supply valve** that can be turned fully off or fully on for heat supply. They also have an **air valve** to vent air at the radiator.

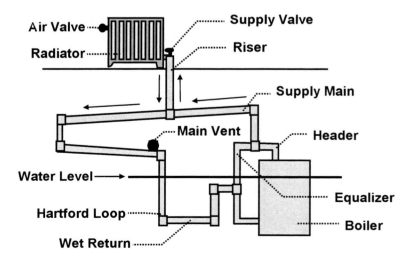

- **2-pipe system:** Radiators in a 2-pipe system (shown below) have 2 pipes, one at each end, to receive steam and to return condensate to the return piping. These radiators also have a supply valve for controlling steam flow, and in a 2-pipe system can be set partially open or closed to control heat. Two pipe radiators don't have an air valve. Instead, the return piping has a **steam trap** which allows air and condensate to pass in the return piping but closes on steam contact. The air is vented through the **main vent**.

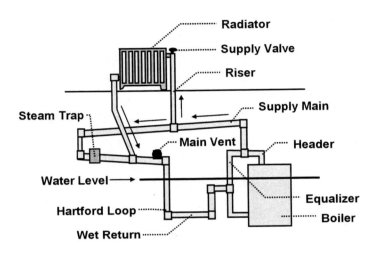

The piping in both 1-pipe and 2-pipe systems must slope downward about 1" for every 10' to 20' of length. That way, condensate anywhere in the distribution piping will flow back to the boiler.

A system is called a **dry return** if the return piping joins the boiler above the boiler water level. A **wet return** joins the boiler below the water level. When a wet return is used, there should be a **Hartford loop** present. The purpose of the Hartford loop is to prevent water from draining out of the boiler in case of a leak in the return line. Water will drain only until it reaches the level of the top of the Hartford loop. A connection between the header at the top of the boiler and the return at the bottom of the boiler is called an **equalizer**. Its purpose is to prevent rising steam from forcing boiler water back up the returns. The equalizer ensures that pressures above and below the water line are the same.

There are variations on the gravity steam system such as the **vapor steam system**, also a 2-pipe system. No air is allowed into this system, and no air needs to be released. Vapor system radiators have a vacuum air valve that prevents air from entering the system.

Definitions

In a <u>gravity steam heating system</u>, water boiled in a boiler changes to steam which rises naturally through pipes to radiators and condensate drains back to the boiler.

A <u>1-pipe steam system</u> has a single pipe to and from each radiator. A <u>2-pipe system</u> has 2 pipes at each radiator and separate return piping.

A <u>Hartford loop</u> is a loop of return piping that turns upwards to connect to the equalizer just below the water level, preventing water from flowing out of the boiler in case of a leak in the return piping. The <u>equalizer</u> is piping from the top of the boiler to the bottom that balances pressure above and below the water line.

**BOILER CONTROLS
AND GAUGES**

- Thermostat

- Limit control

- Pressure relief valve

- Temperature, pressure
  gauge

- Low water cut-off

- Water level sight gauge

- Water fill valve

Steam Boiler Controls and Gauges

A gravity steam heating system is simple and does not require motors or electrical connections other than those for the thermostat and the burner controls. These are the basic controls on a steam boiler:

- **Basic operating controls:** With a steam boiler, the **thermostat** turns on the burners after a call for heat. As steam rises through the system, pressure begins to rise. When operating pressure reaches between 2 psi and 3 psi, a **pressure-sensitive switch** on the boiler will turn off the burners. Some steam boilers are operated with a heat timer instead of a thermostat. The heat timer turns on the burner for a preset number of minutes on the hour or half hour.

- **Limit control:** A steam boiler has a safety device called the high pressure limit control that will turn off the burners if pressure within the system reaches a preset pressure limit of **5 psi**. This control is connected to the boiler with a pigtail pipe (one that has a curl in it). The loop of the pigtail has water in it to prevent the corrosive action of the steam from affecting the control.

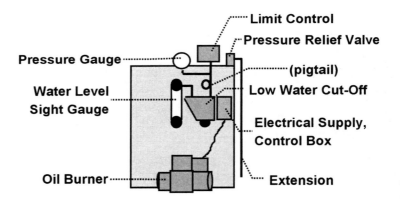

- **Pressure relief valve:** Another safety control is the pressure relief valve that prevents pressure from building to a dangerous level in the boiler. If pressures reach a preset high limit of **15 psi**, the valve will discharge. There should be an extension from the valve to within 6" of the floor.

- **Temperature, pressure gauge:** The boiler will have a gauge showing temperature and pressure readings for the system. If the gauge shows a pressure reading of 15 psi and the relief valve is not discharging, the boiler should be turned off.

- **Low water cut-off:** This safety control shuts down the burner when the level of water in the boiler drops below the designed level. The low water cut-off may be in the boiler or mounted on the outside. The external unit, located in a triangular housing, has a float that drops when the water level falls and turns off the burner. It also has a **blow-off valve** at the bottom. This valve should be opened once a month to remove accumulated silt and mud from the unit which could block the float from dropping.

- **Water level sight gauge:** Steam boilers have a glass tube sight gauge that shows the water level in the boiler. Water levels should vary from 1/2 to 3/4 full — highest when the boiler is cold, lowering slightly when the system is up to pressure and steaming. If the gauge is filled with water or water comes out when opening the top valve on the glass, the system has too much water. If water is not visible in the glass and no water comes out when opening the bottom valve on the glass, the system has too little water.

- **Water fill valve:** Some steam boilers have an **automatic** fill valve that will feed water to the boiler to maintain water levels. This automatic valve may be in a combination unit with the low water cut-off. However, many steam boilers have a **manual** fill valve that must be opened to introduce more water to the system.

Inspecting Steam Boilers

We're not going to repeat all the procedures for inspecting steam boilers here, just those items that set steam boilers apart from hot water boilers. See these pages for other items to be inspected:

- **Thermostats and clearances from combustibles:** pages 9 to 14
- **Gas and oil burners:** pages 24 to 31 and 39 to 43
- **Boilers:** pages 90 to 93

The home inspector should watch for the following conditions with steam boilers:

HOT WATER AND STEAM SETTINGS

A steam boiler has an operating pressure of 2 to 3 psi. A pressure limit control turns off the burners at about 5 psi. Its relief valve discharges at 15 psi.

A hot water boiler has an operating pressure at about 12 psi. A temperature limit control turns off the burners at about 200°. Its pressure relief valve discharges at 30 psi.

For Beginning Inspectors

If you have friends with older homes that have steam boilers, by all means take the time to inspect the boilers. Don't be surprised if you find new replacement parts brought in to keep an old steam system going.

INSPECTING STEAM BOILERS

- Improperly sloped or noisy piping
- Pressure too high
- Inappropriate or fluctuating water levels
- Rusty or dirty water
- Malfunctioning low water cut-off
- Missing Hartford loop

- **Improperly sloped or noisy piping:** As shown in the illustrations on pages 96 and 97, all steam systems should have sloping piping to allow condensate to run back to the boiler. An acceptable slope is 1" for every 10' to 20' of piping. If the slope is wrong and condensate pools in the piping, you'll hear a loud bang or knock as steam hits it at the start of each heating cycle. Inappropriate water levels, clogged air valves on 1-pipe system radiators (sometimes painted closed by the homeowner), or a malfunctioning steam vent on 2-pipe systems can also cause loud banging sounds.

- **Pressure too high:** While the boiler is operating, watch the temperature, pressure gauge. Operating pressures should be within the 2 psi to 5 psi range. If the pressure reading is 15 psi or more, *turn off the boiler immediately* and report the condition as a **safety hazard**. Be sure to inform the homeowner to have the boiler inspected by a qualified technician before re-firing the unit. The cause of the problem may be a faulty pressure relief valve that should be discharging water. During your reading, note if the gauge itself is broken, cracked, and so on.

- **Inappropriate or fluctuating water levels:** The home inspector should pay close attention to the water level sight gauge. CAUTION: If no water is visible in the sight gauge, *do not fire up the system* or if already fired, *turn it off.* Without the appropriate water levels, the boiler can crack as it fires and overheats. The low water cut-off should not be allowing the burners to fire if the water level is too low. A gauge filled to the top indicates too much water in the system. With too much water, it's possible to flood the entire system with water including the piping and the radiators. Both of these conditions should be reported.

The water level in the sight gauge should not fluctuate wildly during boiler operation. As stated earlier, water levels should vary only between 1/2 and 3/4 full — highest when the boiler is cold and lowest when the system is up to pressure and steaming. Rapidly fluctuating water levels can indicate excessive dirt buildup in the boiler or that the boiler is operating at an excessive output. The condition should be checked by a qualified technician.

The glass gauge may show a longtime water level, no matter where the level is at the time of inspection. A high water level and a low rust line in the gauge may indicate that the system is constantly refilling due to a leaky return.

- **Rusty water:** A steam system should be cleaned every few years to remove accumulated rust and mud buildup. Excessive buildup can be seen as rusty or dirty water in the sight gauge. Inform the customer that the system should be flushed by a qualified technician.

  Sometimes, water within the system has been treated to dissolve buildup. It may also be treated with a leak stopper whose granules slowly deposit and plug leaks in the system. If the water in the glass gauge sparkles, a leak stopper may have been added. If you suspect the water may have been treated, let the customer know that it should be tested periodically.

- **Malfunctioning low water cut-off:** During the inspection, you should test the blow-off valve on the low water cut-off. You want to be sure that it will turn off the burner when the water level in the chamber falls. The blow-off valve should be opened once a month to remove silt and mud that can stop the cut-off from operating properly, but some homeowners never do this.

  While the system is operating, open the blow-off valve. Put a bucket under it and stand back as you do so. Rusty, dirty water can spray out and stain your clothes and shoes. And if the valve hasn't been opened for years, mud can come out. As the low water cut-off chamber empties, the burners should shut off. If the burners do not shut off, the unit should be repaired or replaced. Be sure to close the blow-off valve again after this test.

- **Missing Hartford loop:** If the boiler has a wet return (return piping entering the boiler below the water line), there should be a Hartford loop as described on page 97. The Hartford loop prevents water from draining out of the boiler in case of a leak in the return line. Recommend installing a Hartford loop if the wet return doesn't have one.

WHAT'S THE SYSTEM?

If a boiler has an operating water level sight gauge, you've found a steam boiler. (But watch out. Some steam systems can be converted to hot water without removing the old sight gauge.)

DON'T EVER MISS
• Pressure too high
• Low water levels
• Dirt buildup
• Malfunctioning low water cut-off
• Plus all the don't-ever-misses listed for boilers on page 95

REMEMBER: When you finish your boiler inspection, be sure to turn the boiler back on and return thermostat settings to their original settings.

Reporting Your Findings

Follow the directions given on pages 93-95 for reporting your findings for boiler inspections in your inspection report. Be sure to report the type of heating system as a steam boiler

Pay attention to earlier instructions for the reporting of major repairs, safety hazards, and items requiring replacement as stated on pages 94 and 95. These items refer to steam boilers too.

WORKSHEET

Test yourself on the following questions.
Answers appear on page 104.

1. What type of heating system would be indicated if a boiler has an expansion tank but no circulating pump?

   A. Gravity hot water
   B. Forced hot water
   C. Steam

2. What type of heating system would be indicated if a boiler has a water level sight gauge but no expansion tank?

   A. Gravity hot water
   B. Forced hot water
   C. Steam

3. Which heating system is also called hydronic?

   A. Gravity hot water
   B. Forced hot water
   C. Steam

4. What is the normal operating pressure of a gravity hot water system with an open expansion tank?

   A. 2 psi to 3 psi
   B. 12 psi to 15 psi
   C. 28 psi to 30 psi
   D. Water in an open system is not under pressure.

5. The operating pressure in a forced hot water system for a typical 2-story home should be:

   A. 5 psi
   B. 12 psi
   C. 30 psi

6. Pressure in a forced hot water system is maintained by an automatic fill valve and the pressure relief valve.

   A. True
   B. False

7. What condition can be an indication of a defective heat exchanger in a boiler?

   A. Rusting in the burner area
   B. Corrosion in the burner area
   C. Water seepage into the burner area
   D. All of the above

8. What is the purpose of the limit control on a hydronic boiler?

   A. To turn off the burners if temperatures exceed 200°
   B. To discharge water if pressures reach 30 psi
   C. To add water if pressures fall below 12 psi
   D. To shut off the burners if the water level is too low

9. What condition on a hydronic boiler may cause the pressure relief valve to discharge whenever the boiler is on?

   A. A leak in the return piping
   B. A waterlogged expansion tank
   C. An open expansion tank in the attic
   D. A missing relief valve extension

10. What type of boiler is shown below?

    A. A pulse boiler
    B. A condensing hydronic boiler
    C. A steam boiler
    D. A gravity hot water boiler

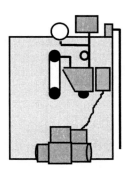

Chapter Nine

OTHER HEATING SYSTEMS

A home may be heated with systems other than gravity or forced warm air furnaces and gravity hot water, forced hot water, and steam boilers.

Electric Heating

Electricity is a clean energy source, but it is usually more expensive to use than gas or oil. Furnaces and boilers may be powered by electricity, which has already been discussed on pages 58 and 83. There are other ways electricity can be used to heat the home:

- **Resistance heating:** Electric baseboard convectors, wall mounted strips, or floor inserts may be used to heat the home, but this method is not to be considered a central heating system. Each resistance unit in the house, one per room most likely, operates as a separate heating plant. Each resistance heater has its own manual thermostat, either on the unit or wall mounted. The wall mounted strips and floor inserts usually have a fan to circulate the warm air.

  Resistance heaters can get very hot, so their clearance from combustibles is important. Curtains and draperies should be kept 8" above the heaters or if hanging in front of the heaters should be 3" away and 1" above the floor.

- **Radiant heating:** Similar in concept to hot water radiant heating, electric radiant heating consists of electrical cables embedded in ceilings or floors during construction. The cables heat the surface, radiating heat into the room. Each room would have a separate control. If the distribution wiring breaks, it can be difficult to locate the problem. Most often, when electrical radiant heating fails, the system is abandoned and replaced with electric baseboard heaters.

- **Heat pumps:** Electric powered, a heat pump is a reverse-cycle air conditioning system. In the air-to-air heat pump, cold air is sent into to the house and warm air expelled outdoors when in the air conditioning mode. In the heating mode, cold air is expelled outdoors while heat is sent into the house. This will be discussed more fully later in this guide, beginning on page 133.

Worksheet Answers *(page 103)*

1. A
2. C
3. B
4. D
5. B
6. A
7. D
8. A
9. B
10. C

Other Heating Systems

The home inspector may find **combination heating systems**. For example, one zone of the home may have forced warm air heat while another may still be using a gravity hot water system. Any combination may be possible. In a case like this, both systems should be inspected.

There are also **hydro-air systems** that combine hot water and forced air. The way this type of system operates is that hot water is heated in a boiler and sent to a hot water heat exchanger coil in a blower unit. The hot water coil heats the surrounding air which is blown into the home through normal supply ducts.

The home inspector should not be upset when coming across a heating system that he or she hasn't seen before. Take the time to examine the system and try to figure out how it operates. Many times, unfamiliar systems will have familiar components that you've seen before. Don't be afraid to let the customer know that you've found an uncommon system and that you may not be able to tell them everything they need to know about it. It's best to be honest with customers up front. Suggest that a qualified technician come in to inspect the system.

ELECTRIC HEATING

- Furnaces and boilers
- Baseboard, wall, or floor resistance heating
- Ceiling or floor radiant panel heating
- Heat pumps

Wall Furnaces – Usually found in smaller homes or cottages, or installed in added rooms. Older unit may be unsafe. Do Not attempt to light the pilot. Clearances are 12" from a door, 6" from a wall, and 18" from a ceiling.

Floor Furnaces – The floor furnace is similar to the wall furnace, except that it is suspended between floor joists. Lift the grille, if possible, and look for deterioration

Pages 106 to 112 outline the content and scope of the cooling inspection. It's an overview of the inspection, including what to observe, what to describe, and what specific actions to take during the inspection. Study these guidelines carefully. This section also mentions some special rules about when and when not to operate the cooling system.

Chapter Ten

THE COOLING INSPECTION

This chapter will present information on central air conditioning systems.

Inspection Guidelines and Overview

These are the standards of practice that govern the inspection of the home's cooling system, and by that we mean central air conditioning. Please study these standards of practice carefully.

Cooling System	
OBJECTIVE	To identify major deficiencies in the central air conditioning system.
OBSERVATION	Required to inspect and report: • Central air conditioning • Through-wall Units — Cooling and air handling equipment — Normal operating controls • Distribution systems — Fans, ducts and piping, with supports, dampers, insulation, air filters, registers, radiators, and fan-coil units • The presence of an installed cooling source in each room Not required to observe: • Window air conditioners • The uniformity or adequacy of cool-air supply to the various rooms
DESCRIPTION ACTION	Required to Describe: • Energy sources • Cooling equipment type Required to: • Operate the system using normal operating controls. • Open readily openable access panels provided by the manufacturer or installer. Not required to: • Operate cooling systems when weather conditions or other circumstances may cause equipment damage.

For Your Library

Home Heating and Air Conditioning Systems by James Kittle and Preston's Guide... are both helpful to the air conditioning inspections. The Preston's Guide is a good tool to have for determining year of manufacture and remaining lifetime. Add them to your library. Call 800-441-9411 to get information on ordering.

The standards provide an outline of what is to be inspected and what is to be reported during the cooling inspection. Here's more of an overview:

- **Cooling equipment source**. The inspector is required to **open readily openable manufacturer's access panels** during the visual inspection and to **operate the cooling system using normal operating controls**.

  The home inspector is not required to operate the air conditioning system when weather conditions don't permit. A general rule for home inspectors is to not turn on the cooling system if the outside temperature is less than 60° at the time of the inspection or dropped below 60° the night before. Also, the home inspector is not required to operate the cooling system if the system is shut down, has not yet been operating this season, or if the circuit breaker for the unit is turned off.

  The **condition** of the compressor, condenser, evaporator, air handler, and connections are all inspected. The inspector examines each, looking for defective operation, dirt, noise, and leaking. The inspector also determines the **age** of the cooling unit and its remaining useful lifetime. Most cooling systems, primarily the compressor last about 8 to 12 years. In the sunbelt where these systems are used more frequently, the lifetime would be closer to 8 to 10 years. Any central air conditioning system over 10 years old should be reported as **an item needing replacement within 5 years**.

  The home inspector is **not required to inspect non-central air conditioners** such as window units, but is typically required to inspect wall units.

- **Distribution system:** The cooling system's distribution system may be the same as that for the forced warm air furnace, including the fan, filter, and ductwork. These items are normally inspected as part of the heating inspection. But for independent, standalone cooling units, the home inspector inspects the fan, ductwork, drain piping, air filters, and other items for their condition. Defects such as restricted air flow, loose ducts, and vibrations are all reported.

THE COOLING INSPECTION

- Cooling equipment and operating controls
- Distribution system
- Cool-air source per room

- **Cooling source per room:** An easy part of the inspection, but an important one, is to check for the presence of a cool-air source in each room. The inspector records the presence or absence in the inspection report. However, the inspector is **not required to observe the uniformity or adequacy of the cool-air supply to each room**.

NOTE: Effective January 23, 2006, residential cooling units sold must be 13 SEER or higher. SEER stands for Seasonal Energy Efficiency Ratio, the method used to judge how efficiently an air conditions performs.

The new 13 SEER requirement does not mean you have to replace your cooling system now. But when you do decide to replace your cooling unit, installing a unit rated 13 SEER or higher provides maximum efficiency.

The Refrigerant Cycle

An air conditioning system has 3 basic components which are the **compressor**, a **condenser**, and an **evaporator**. These components are connected in a sealed system. The sealed system contains a **refrigerant** that changes from liquid to gas and back to liquid as it flows through the system. The refrigerant in home cooling systems is a chlorofluorocarbon, commonly called **Freon**. The system also includes a means of removing heat from the condenser and a means of circulating air across the evaporator and through the house.

The operating principles behind the refrigerant cycle in air cooling system are the following:

1. **Gas heats up as its pressure is raised.** The **compressor** increases the pressure of the refrigerant when it is in its gas state. This causes its temperature to rise. That's so it is capable of giving off heat.

2. **Heat is removed from the gas, causing the gas to become a liquid.** As the pressurized gas moves through the **condenser coils**, heat is removed by means of a fan (air cooled) or by means of water (water cooled). When heat is removed, the gas condenses to a warm liquid that is under a high pressure.

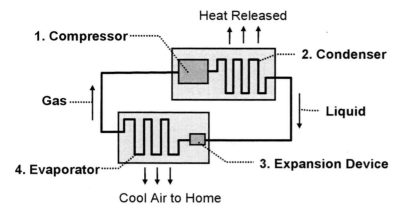

1. Compressor
2. Condenser
Heat Released
Gas
Liquid
4. Evaporator
3. Expansion Device
Cool Air to Home

3. **Liquid cools as its pressure drops.** The high-pressure liquid passes through an **expansion device**, or pressure-reducing device, causing its temperature to drop below room temperature.

4. **The liquid absorbs heat and turns back into a gas.** The liquid refrigerant passes through the **evaporator coils** and absorbs heat from the air passing over the coil, thereby cooling the air. A fan circulates the cool air to the house. As the liquid absorbs heat, it turns back into a cool, low-pressure gas and the process begins again.

Inspection Concerns

The main concern the home inspector should have during the inspection of the cooling system is not to damage the equipment. Operating the system under certain conditions can cause damage to the compressor. The following rules should be followed in order to avoid doing harm to the unit. First, **do not operate** the cooling equipment if:

- **The outside temperature is less than 60°** or if the temperature dropped below 60° the night before the inspection. Most manufacturers suggest this 60° mark for safe operation. If the outside temperature is right at 60° and you're unsure whether to start the equipment, you can touch the bottom of the compressor shell if it's accessible. It should be warm to the touch. If it's cold, you shouldn't turn the system on.

  For a **heat pump**, the unit should not be operated in a cooling mode if the temperature is **less than 65°** or in a heating mode if the temperature is over 65°. (We'll talk more about heat pumps later in this guide.) The home inspector will inspect the heat pump either in its cooling mode or its heating mode, not both. If the unit is already operating in the cooling mode, you would inspect its cooling mode. If the unit is already in the heating mode, you would inspect its heating mode.

- **The unit is shut down at its main power source.** Ask the homeowner why the system is off and then decide whether it is safe to operate the equipment without damaging it. *It may not be.*

The compressor's motor, safety devices, and oil lubricant are sealed in a shell. When the system is off, the oil can absorb the refrigerant so it can no longer protect the moving parts. Most compressors have a heater to maintain a high enough temperature to prevent this absorption. However, if the circuit breaker is turned off to the cooling equipment, it's also off to the heater. If you start up a cold system, you can cause damage since the oil will no longer be able to lubricate properly. The system's power supply should be turned on for **24 hours before startup** to allow the heater time to perform its work. Some home inspectors finding this condition will turn on the power and come back the next day to inspect the cooling system. Or you may want to call the day before the inspection to have the power turned on.

- **The unit hasn't been operated this season.** For the same reasons given above, be careful about operating the cooling equipment if it hasn't yet been used this season. Check first to see if the power has been on to the unit for at least 24 hours and the heater is operating. If the power is still off, follow the directions given above.

NOTE: If you've just inspected the furnace and the cooling system's evaporator coil is located in the furnace plenum, wait 5 minutes after the circulating fan stops to give the plenum time to cool down before starting the air conditioning unit.

The other concern of the home inspector while inspecting the cooling system is when to turn off the system if something appears to be going wrong. There are certain signs that the equipment is malfunctioning. The home inspector **should recommend servicing** if during its operation:

- **The condenser fan or the compressor goes on but not both.** The condenser fan and the compressor should both start upon a call for cool air at the thermostat. If only one of them goes on, it may be that you've activated a time delay on the compressor. Some compressors have a time delay at 2, 3, or 7 minutes. Wait to see if the compressor will start up before you decide if there's a problem. If the compressor doesn't start, something is wrong with the equipment, in which case the system should not be allowed to continue operation.

- **The compressor groans or squeals.** The compressor should run smoothly, so any unusual noises indicate a problem condition. The equipment should be turned off and examined by a serviceperson.

- **The compressor short cycles.** The compressor should go off and on in response to the thermostat, not on and off repeatedly. This condition indicates a problem and the unit should be shut down.

- **Any part of the cooling system is not operating.** If, during your inspection of the cooling system, you notice any part of it not operating even though the power is on, shut down the system before damage is done.

If you don't operate the air conditioning system for any of the reasons given above or have to turn off the system once started, be sure to share your reasons with the customer. Explain the situation and why the unit can't be inspected. In our northern climate, of course, we don't have the opportunity to inspect the cooling system during the winter and a good deal of the spring and fall. Our customers understand the situation. They would rather have the system inspected later than to have damaged equipment. When we don't operate the system due to some malfunction, we always recommend the system be checked out by a qualified serviceperson to find the problem.

A Word about Cooling Capacity

It's possible to tell if a cooling system has the correct capacity for the house it serves. However, the inspector is not required to judge the adequacy of the cooling capacity or report this fact to customers. The information given here is for your own information.

Personal Note

"I don't mean to scare you with these warnings, but it is important not to start or continue to operate air conditioning equipment if there's a chance you may damage it. If the basic rules as laid out on these pages are followed, you should be all right."

Roy Newcomer

RATINGS IN TONS

A ton of air conditioning will provide 12,000 BTU's of cooling per hour. The amount of cooling capacity varies depending on your region of the country.

Cooling systems may be rated in **either BTU's or tons**. A ton of air conditioning will provide 12,000 BTU's of cooling an hour. (When referring to cooling capacity, 1 BTU equals the amount of heat that must be absorbed to *lower* the temperature of 1 pound of water 1° Fahrenheit.)

A general rule to follow is that 1 ton of cooling capacity is needed for every 400 to 800 square feet in the home. This is only a general rule, however. It may be that 1 ton of cooling is sufficient for every 700 to 1000 square feet in moderate climates, while warmer climates may require 1 ton of cooling for every 450 to 600. These capacities may also vary due to differences in size and layout of the house, the amount of insulation, and so on. Only an air-conditioning professional can make an accurate determination of needed cooling capacity.

Another reason why home inspectors are not required to determine the adequacy of cooling power is that it's difficult to determine what the tonnage or BTU rating is for the cooling system. Manufacturers tend to state tonnage or BTU ratings within the model or serial number of the unit. However, they use so many different codes, it's hard to know what the numbers mean. For example, in a Trane unit, the first digit in a certain model number indicates tonnage — 3XXX would indicate 3 tons. But numbers in a Lennox unit indicate BTU — HSX-311 indicates 31,100 BTU's. It's very confusing, and The Preston's Guide does not help you sort out all model codes.

Chapter Eleven

AIR COOLED AIR CONDITIONING

Air cooled air conditioning systems can work in **conjunction with a forced warm air furnace**, where the furnace fan and duct system are used by the cooling system. Air cooled systems may also operate as **standalone units** (if the house has a boiler, for example, not a furnace) with its own fan and its own ductwork.

The Furnace-Related Cooling System

This type of cooling system has 2 main components. An **outdoor component** houses the compressor, the condenser, and a fan. The refrigerant in a gaseous state is compressed by the **compressor**. As the gas runs through **condenser coils**, heat is removed from the gas by a **fan** that pulls outside air across the coils. Then the gas loses heat and condenses to a warm, high-pressure liquid state.

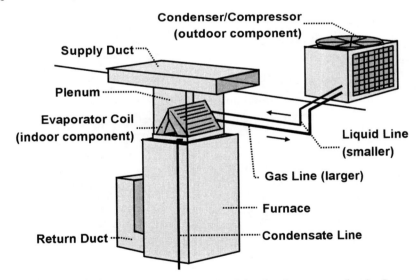

The liquid refrigerant is sent inside the house to the **indoor component**, which sits in the plenum directly above the furnace. Here, an **expansion device** reduces pressure on the liquid which cools it down. The liquid flows through the **evaporator coils**, where it absorbs heat from the air flowing through the plenum. This cools the surrounding air, and the furnace fan circulates it to the house through the furnace ductwork. As the liquid picks up heat again, it condenses back in to a gas. The cool, low-pressure gas returns to the outdoor unit. The evaporator sits in a

COMPONENTS

- Outdoor unit with the compressor, condenser coils, and fan

- Lines for gas refrigerant and liquid refrigerant

- Indoor plenum unit with an expansion device, the evaporator coils, condensate tray, and condensate line

- Furnace components including fan and ducts

CLEARANCES

- 12" on all sides of the outdoor cabinet

- 4' to 6' above the outdoor cabinet

condensate tray or pan that catches condensation that forms on the outside of the cold coils. Condensate is discharged through a **condensate line** to a floor drain in basement units, or to the exterior in attic mounted units.

Here is a more detailed explanation of the components of an air cooled air conditioning system that works in conjunction with the forced warm air furnace:

- **Outdoor cabinet:** The outdoor cabinet houses the compressor, the condenser, and the fan. It is a large box-like or cylindrical affair with louvered sides. There should be at least **12" of side clearance** from foliage or obstruction and **4' to 6' above it**. And the cabinet should be located where there is a minimum of direct sunshine, since the cooler the air flowing over the condenser, the more efficient the refrigerant cycle will be. The cabinet should be **level**, mounted on a concrete pad to which it is securely fastened, and high enough to be out of the way of rain and snow. A tilt of 10° can damage the compressor.

There should be an **exterior electrical disconnect switch** near the outdoor cabinet so the unit can be turned off for maintenance. The unit is on a 240-volt circuit with fuses or circuit breakers. During the off-season, the circuit breaker should be turned off to prevent any inadvertent action at the thermostat from starting the cooling system.

OUTDOOR COMPONENTS

- Outdoor cabinet
- Compressor
- Condenser
- Condenser fan
- Disconnect switch

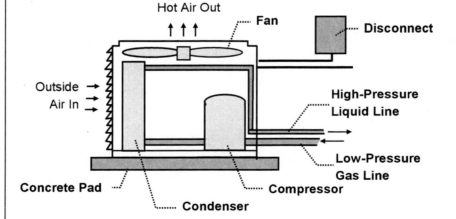

#28 Outdoor compressor/condenser cabinet

*Photo #28 shows an **outdoor compressor/condenser cabinet**. Note its location in a level open area, free of obstructions for good air flow. The unit should be kept clean as shown here, with no debris allowed to accumulate on top or at the side louvers.*

- **The compressor:** The compressor is the heart of the cooling system and its most expensive component, costing about 30% to 50% of the entire system. It typically lasts from 10 to 15 years, or perhaps 8 to 12 years in the sunbelt. When the compressor wears out and has to be replaced, it may be more cost effective to put in a whole new air conditioning system depending on efficiency and age (see 13 SEER note).

The **function** of the compressor is to move the refrigerant through the system and to compress refrigerant gas until it becomes a high-pressure, high-temperature gas capable of giving off heat. Low-pressure gas is fed to the compressor through the low-pressure refrigerant line from the evaporator, and high-pressure gas is delivered from the compressor to the condenser coils.

The compressor is a **sealed unit** with a motor, safety devices, and an oil lubricant. (It's not possible for the home inspector to inspect the compressor other than to determine whether it is operating properly.) When the cooling system is shut down, the oil will slowly absorb the refrigerant and no longer be able to protect the moving parts of the compressor. To prevent this condition, most compressors have a **heating element** that keeps the oil free of refrigerant. The heater will continue to work as long as the power supply to the system is not turned off.

Definition

The <u>compressor</u> moves the refrigerant through a cooling system and pressurizes the refrigerant gas in order to raise its temperature.

REFRIGERANT LINES

- The <u>larger line</u> carries low-pressure gas from inside to outside. It should be <u>cool</u> to the touch.

- The <u>smaller line</u> carries high-pressure liquid from outside to inside. It should be <u>warm</u> to the touch.

Definitions

The <u>condenser</u> in a cooling system is a coil through which the refrigerant gas flows, releasing heat to the air and becoming a liquid.

<u>Refrigerant lines</u> are copper piping that carry refrigerant in both gas and liquid states through a cooling system.

Before restarting the cooling system at the beginning of the season, power should be turned on **24 hours before startup** to allow the heater time to heat the oil. Remember also that the outside temperature should be **above 60°** at the time of startup and not fallen below 60° the night before. Starting the system under the wrong conditions can severely damage the compressor.

The compressor goes on and off in response to the thermostat. Upon a call for cool air at the thermostat, the compressor and the fan in the outdoor unit should both start up. The compressor should run smoothly and quietly, without going on and off repeatedly. The bottom of the compressor should feel warm or hot to the touch. If it feels cold, there's something wrong.

- **The condenser:** The condenser is a passive, non-mechanical device. It consists of a copper or aluminum coil of thin wall tubing covered with fins to increase its surface area. The function of the condenser is that of a **heat exchanger**. Outside air is pulled over the condenser fins, transferring heat from the high-pressure refrigerant gas inside the condenser to the air. This causes the high-pressure refrigerant gas to condense to a warm liquid. The condenser fins should be kept clean, unobstructed, and free of damage.

- **The condenser fan:** The fan in the outdoor cabinet moves air over the condenser coils. Warm outside air is pulled into the cabinet to help cool the refrigerant in the condenser. Hot air is expelled back to the outdoors. Fans may be placed vertically or horizontally in the cabinet. They should be kept clean and undamaged.

- **Refrigerant lines:** There are 2 copper lines that move the refrigerant through the cooling system. The **low-pressure line** carries the cool low-pressure refrigerant gas from the evaporator to the compressor (from inside to outside). It's about the width of a broom handle and should be insulated to prevent condensation from forming. When the system is operating properly, the low-pressure line should be cool to the touch. The **high-pressure line** carries the warm high-pressure refrigerant liquid from the condenser to the evaporator (from outside to inside). This line is thinner like a pencil and feels warm or hot.

#29 Refrigerant lines

Photo #29 shows the refrigerant lines behind the compressor/condenser unit. These are the 2 lowest lines at the house connection (the other 2 lines from the electrical box are electrical lines). Notice the fatter low-pressure gas line (should feel cold) and the much thinner high-pressure liquid line (should feel warm). These lines should be checked for any kinks, twists, or damage. If the fatter line is frosted, it can indicate that the system is low in refrigerant. A sight glass may be mounted in the thinner line. Bubbles seen through the sight glass may also be an indication that the refrigerant is low.

- **Indoor unit:** The indoor unit that works in conjunction with the furnace sits in the plenum above the furnace. Its components include the evaporator coil, the condensate tray, and the condensate line.

- **The evaporator:** The evaporator is like the condenser in configuration in that it has copper or aluminum tubing to which thin fins are attached. A common evaporator coil is called an **A-coil** because of the double plates of coils connected at the top. The liquid refrigerant first flows through an **expansion device** to drop its temperature and then flows through the evaporator coils as the furnace fan blows air across them. The liquid absorbs heat from the warm air and cools the tubing and fins which likewise cool the air. The cool air is circulated to the house through the furnace ductwork. The air is also dehumidified in the process. As the refrigerant absorbs this heat, it changes back to a low- pressure gas.

When the evaporator coils sit in the plenum, they're not typically visible to the home inspector, but you'll know it's there because of the refrigerant lines at the plenum.

Definition

The evaporator in a cooling system is a coil through which the refrigerant liquid flows, absorbing heat from the air and becoming a gas.

INDOOR COMPONENTS

- The expansion device
- The evaporator coils
- The condensate tray
- The condensate line
- Furnace fan, filter, and ductwork

Photo #30 shows an A-coil in the plenum. Here, we had to remove the humidifier cover to get a photo of it (we don't suggest you do that during the inspection).

#30 A-coil in the plenum

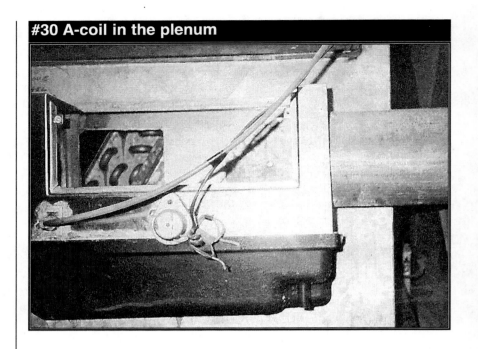

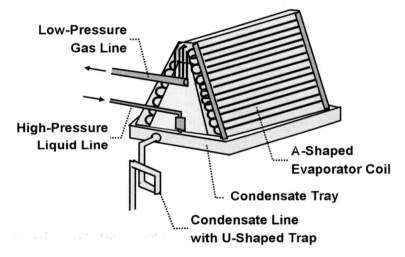

Low-Pressure Gas Line

High-Pressure Liquid Line

A-Shaped Evaporator Coil

Condensate Tray

Condensate Line with U-Shaped Trap

The **temperature differential or split** of the air before and after the evaporator should be between 15° and 22°. If the air on the outlet side is too warm, it can be the result of too little refrigerant in the system, too much air passing over the evaporator, or too high fan speeds. If the outlet air is too cold, it can be the result of a dirty filter, clogged evaporator coils, ice on the coils, or too low fan speeds.

- **Furnace fan, filter, and ductwork:** Normally, a furnace fan is a 1-speed fan, not adequate for use with a cooling system unless it's modified for multiple speeds to increase air flow for the cooling system. Frosting or icing coils are an indication that air flow isn't adequate. The furnace air filter should be changed every month during the heating and cooling seasons to keep the evaporator fins clean.

The cooling system uses the furnace ductwork for distributing cool air through the house. More duct capacity is needed to circulate cool air than warm air. The heavier cool air, blown at greater speeds, drags along the ducts with more friction. If a home has small 4" round heating ducts, it's likely that the cooling power of the air conditioner will be reduced. These undersized ducts can whistle and vibrate as the cool air is forced through them.

Ideally, air conditioning registers should be located in or near the ceiling. However, that's not how furnace ductwork is normally installed. However, some homes with furnaces have **high-low registers**. In this system, high registers for cooling and low registers for heating are present in each room. These combination register systems have to be balanced by the register dampers at each change of season — 2/3 of the supply capacity should be high in summer and 2/3 should be low in winter.

Any ductwork in the attic or crawl space should be insulated to reduce the effect on the cooling system of high summer temperatures in these areas.

NOTE: When the furnace has a **humidifier** that has a bypass from its location in the return duct to the plenum, air flows through the bypass to the plenum. During the cooling season, a damper in the bypass should be closed down. If there's no damper present, air flowing through the bypass can affect the air conditioner's efficiency and may ice up the evaporator coils.

- **The condensate tray:** When household air passes over the evaporator, condensation forms on the coils. This condensation is collected in a condensate tray under the coil where it can be removed. This tray should be free of leaks, cracks, rust, and obstructions so water can drain properly. Some of the old air conditioning systems had tin pans that rusted out after much use. Because of the coil's position above the furnace heat exchanger, leaks or water overflowing from the condensate tray can drip onto the furnace heat exchanger.

- **The condensate line:** The condensate drain line should have a trap in it to prevent cool air escaping through the pipe. The line should run to a floor drain, to a nearby sink, through the foundation andonto ground outside the house or as appropriate for your region. Condensate is not drinkable, and it should not be allowed to enter the domestic water supply. You may find the line terminating in a hole in the slab allowing the condensate to drain into the ground under the slab. This is not an acceptable practice.

Occasionally, you may find the condensate line discharging into a small rectangular box next to the furnace. This box contains a **lift pump** that pumps the condensate to a level where it can be sent to a desired location. The pump has a float control that activates the pump when water reaches a preset level.

The Standalone Air Cooled Unit

When a home doesn't have a forced warm air furnace and is heated by hot water or steam, central air conditioning must operate independent of the heating system. These standalone air cool systems are identical in operation to air cooled system described in pages 113 to 120. They have the identical components outdoors in the compressor/condenser cabinet, but the inside components are contained in a **standalone unit or air handler**. This cabinet consists of the evaporator coils, a fan, its own air filter, a condensate tray and line, and connections to its own supply and return ductwork.

*Photo #31 shows a **standalone indoor cabinet**. Here, you can see the squirrel cage fan and fan motor, the evaporator coils, the refrigerant lines, and the condensate tray at the bottom. This unit happened to be located in the basement. Although you may find some standalone cabinets in the basement or even a closet, they're most often mounted in the attic. Since the attic units can cause a lot of noise and vibrations in the house, they should be mounted on rubber, cork, or Styrofoam pads.*

#31 Standalone indoor cabinet

Here are some special considerations for attic mounted standalone units:

- **Condensate drainage:** The base of the standalone cabinet is sometimes the condensate tray too. Its condensate line should have a trap in it near the cabinet. When units are attic mounted, there is concern that a cracked or clogged condensate tray can leak and damage the ceiling below. Some cabinets have a fitting for an **auxiliary condensate line** just above the main line fitting. If the main tray and line become clogged, the level of condensate will rise and be drawn off by the auxiliary drain. If an auxiliary line fitting is missing, the whole cabinet, including the condensate tray, should sit on an **auxiliary condensate drip pan** with its own drain line as a backup.

The condensate lines should drain through the lower portion of the roof to a gutter or through the exterior wall to the outside. The lines should <u>not</u> terminate into the plumbing vent stack. Most local codes do not permit this because of the possibility of sewer gas backing up into the air flow through the fan unit and entering the house.

NOTE: The condensate tray line and the auxiliary drain line should be 2 completely separate lines. It defeats the purpose if the lines are joined.

- **Ductwork:** Since a home without a furnace won't have ductwork for moving air through the home, a standalone air cooled system must have its own ductwork installed. Most of these systems have round, small diameter, flexible, insulated ductwork. The air velocity is increased to allow enough air flow through the small ducts. Instead of using conventional-style heating registers, discharge nozzles are mounted in ceilings to blow in cool air. A large return grill is usually mounted in the ceiling on the top floor.

Definitions

A <u>standalone air cooled air conditioning system</u> has an outdoor compressor/condenser unit and an independent indoor evaporator unit with its own fan, filter, and ductwork. It does not work in conjunction with a forced warm air furnace.

An <u>auxiliary condensate drip pan</u> is used with an attic mounted standalone unit as a backup against leaks in the main condensate tray and line.

DO NOT OPERATE

- If outside temperature is under 60° or under 60° the night before.
- If system is shut down.
- If system hasn't been operated this season.

RECOMMEND UNIT BE SERVICED

- If compressor or condenser fan doesn't start.
- If compressor groans and squeals.
- If compressor short cycles.
- If any part doesn't work.

Inspecting Air Cooled Systems

During the inspection of the air conditioning system, the home inspector should identify the **type** of system and determine its **age** and remaining useful lifetime. The inspector should make a determination as to **whether or not to operate** the system. The inspection should include examining the following air conditioning components:

- Outdoor cabinet
- Refrigerant lines
- Indoor unit
- Ductwork

These procedures should be followed when inspecting the cooling system:

1. **Examine the compressor/condenser unit:** Begin your examination of the cooling system by starting with the outdoor unit while performing the exterior inspection of the home. Note the presence of an electrical disconnect, clearances around the unit, the condition of the cabinet and fan blades, whether the cabinet is level, and whether the refrigerant lines are kinked or damaged.

2. **Determine if you'll turn on the unit:** It's always a good idea to know what the temperature was the night before the inspection. Then when you're outside the home during the inspection, take note of the current temperature. When you're examining the outdoor unit, note if the disconnect switch is turned off, indicating that the system is shut down. Talk to the homeowner and find out why the system is off and if it can be turned on again without damage. Remember that if the unit hasn't been operated yet this season, it should have its power on for 24 hours before the inspection to prevent damage to the compressor.

3. **Turn on the equipment.** If you've decided to go ahead and turn on the equipment, lower the thermostat while you're inspecting the interior of the house. Listen for any problems with the air conditioning as the system starts up. Be sure to shut off the equipment if the compressor and condenser fan don't start up immediately, if you hear problem noises, or if any parts aren't functioning.

4. **Let the equipment run for 15 minutes.** Continue with other aspects of the interior inspection and let the air

conditioning operate for at least 15 minutes. As the cooling system comes up to speed, check registers and return grills for air flow. Check visible ductwork.

5. **Go outside once more to check the outdoor unit.** While the system is still operating, check the outdoor unit once more. Check fan operation and feel for warm air blowing out of the unit. Feel the refrigerant lines for an indication of any problems — the larger line should feel cold and the smaller line warm.

6. **Examine the indoor unit.** Allow the cooling system to continue operating while you listen to the indoor evaporator unit. Check the temperature of air before and after the coil for appropriate differential. Note any excessive vibrations, fan motor problems, condensate leaking, functioning lift pump, and so on.

7. **Turn off the cooling system.** Finally, turn off the system by having someone turn up the thermostat. Then inspect the indoor unit more closely, checking the condition of the cabinet for rusting and corrosion, the presence and condition of the air filter, mounting (if an attic unit), and condition of the condensate tray and line(s).

Watch for the following conditions during the inspection of the air cooled air conditioning system:

- **Obstructed or tilted outdoor unit:** Check the outdoor unit to be sure the proper clearances at the sides and tops are met. Check if the unit is level and explain why a tilt of only 10° can be harmful to the compressor. Be sure to report unlevel compressor/condenser units in your inspection report. Excessive settlement of the concrete slab under the unit can cause refrigerant lines to fracture and lose refrigerant.

- **Missing exterior disconnect switch:** There should be an electrical box near the outside cabinet so the system can be turned off at this point and will not respond to the thermostat during the off season. This is a precaution against someone turning down the thermostat at the wrong time, starting the compressor and damaging it.

- **Damaged, dirty, leaking, corroded, or rusted equipment:** Check out all the outdoor and indoor components of the

*Photo #32 shows an **interesting compressor/condenser unit**. At first glance, we thought that there was snow in the unit. But upon looking closer, we realized that the whole whitish area was corrosion. The cabinet louvers, the condenser coils, and even the fan were in terrible shape. Can you guess the cause? The family dog had marked this as his territory. This was his favorite spot to relieve himself. Of course, being winter, we couldn't operate the air conditioner, but we suggested that a serviceperson be called in the spring to check out the system. (We also suggested that the family have a talk with the dog.)*

cooling system (you may not be able to see the evaporator coils in the plenum). Cabinets and all operating components should be kept clean and free from rust and corrosion. Look for and report dirty, rusted, or clogged coils or broken fins, broken blades on the condenser fan, any evidence of leaking components, and damaged and leaking refrigerant and condensate lines.

Take special care when inspecting **condensate trays** in the plenum since leaking above the heat exchanger can have dire consequences to the condition of the furnace heat exchanger. If you find leaking from the tray into the furnace, take that into consideration during your inspection of the heat exchanger.

#32 Interesting compressor/condenser unit

- **Missing or dirty air filter:** If you find the air filter missing in a cooling system, you should shut the equipment down. You can bet the evaporator coil will be dirty and clogged, and the system isn't operating properly. Air filters are very important to the smooth operation of a cooling system. Remind customers that filters should be cleaned or replaced once a month during the heating and cooling seasons.

- **Noisy, vibrating, or malfunctioning equipment or loose mountings:** Examine as much of the equipment that you have access to for these conditions. The compressor, condenser fan, inside blower unit, and lift pump (if present) should all operate smoothly without excessive noise and vibration. Explain to customers the seriousness of some of the defects you may find. For example, a noisy or short cycling compressor represents a serious problem and expensive repair, while a dirty or vibrating condenser fan can be repaired easily. A serviceperson should be called in to examine any compressor problems. Vibrations can be caused by loose mountings rather than faulty equipment.

- **Frost or icing:** Any sign of frost or ice buildup on the evaporator coils or the refrigerant lines is an indication that the system isn't operating properly. It's usually the result of an insufficient air flow through the coil or not enough refrigerant in the system. This should be reported in the inspection report.

- **Too high or too low temperature differential:** You can note temperature differentials in the cooling system in 3 ways — temperature differences in air flow into and out of the outdoor unit, between the larger and smaller refrigerant lines, and in air flow into and out of the evaporator coil.

A simple check for system operation is to feel if the air blown out of outdoor unit is considerably warmer than the outside air. You can also feel the refrigerant lines — the larger line should feel cold and the smaller warm. If the larger line is not cold, the system is probably low on refrigerant (bubbles in the sight gauge is also an indication of this).

At the evaporator coil, the home inspector should determine if there's a **proper temperature differential of 15° to 20°** between the return and supply side of the evaporator coils. Too low a differential (less than 15°) can indicate the unit is low on refrigerant, has a bad compressor, or is too old to transfer heat properly. Too high (greater than 22°) can indicate a restricted air flow from a dirty evaporator coil or air filter, or dirty or malfunctioning fan. An extreme temperature differential should be reported in your inspection report.

- **Improper condensate drainage:** Always closely examine the condensate tray and condensate lines for their condition, watching for corrosion, rusting, or leaking. Note if a trap is present in the line. Report any improper drainage such as draining the condensate to a hole in the slab or into a plumbing vent. For attic installations, note the absence of an auxiliary drip pan under the air handler and/or the absence of a separate auxiliary drain.

• **Leaking, damaged, loose, or corroded ducts:** Examine the visible ductwork for condition, noting any open joints, corrosion, loose supports, or missing insulation.

• **Lack of cooling source in each room:** If a cool-air source is missing in any room, note its absence in the Inspection Report. This is a similar situation to reporting heat sources. Customers will be calling later wanting you to install a cool-air outlet in any room you missed on the report.

Reporting Your Findings

During the cooling inspection, be sure to continue to communicate with your customer about your findings. Often, customers don't understand why you aren't testing the cooling equipment. Explain your reasons for not operating the system without giving a boring technical lecture, but say enough to help customers understand how equipment can be damaged if the system is started at the wrong time or under the wrong conditions. Customers may not fully understand how the air conditioner works, but they should be able to understand that you are taking measures so as to not damage the equipment.

If a home has a central air conditioner and you find a room that is missing a cool-air register, be sure to make a note of it in your inspection report on the appropriate page for the kitchen, the bathrooms, or any other room. Missing a cooling source in a room can cause as much trouble later as missing a heating source.

When reporting on the inspection of the air conditioning system, be sure to report on the condition of the following:

• **Outdoor unit:** Your inspection report might be organized so that the air conditioning outdoor unit (compressor and condenser) is reported with other exterior items. Record the brand name, model, and serial number of the unit. You should also give the condition of the outdoor unit an overall rating of satisfactory, marginal, or poor, noting defects such as a tilted unit, obstructions to the unit, rusted components, broken fan blades, a noisy compressor, and so on. Make a note indicating whether an electrical disconnect is present.

- **Indoor unit:** You may be reporting information about the indoor unit on a different page of your report. In any case, report on the type of air conditioning system you find and indicate its fuel source (gas or electric, for example). Report on findings you may have such as a leaking condensate tray, missing air filter,

- **Operation:** First of all, be sure to distinguish between an air conditioning system you haven't operated and one that isn't operating due to malfunction. Be clear about reporting this. In other words, note if the unit was or wasn't started due to outside conditions. It's a good idea to actually write the reason why you didn't turn it on (outside temperature too low or power not on for 24 hours, for example). If you tried to start the air conditioner and it didn't run, then note that. Report on whether the operating temperature differential was within range.

- **Major defects and repairs:** If the air conditioning system is not in operating condition, report it as a major defect. Serious compressor problems represent a major repair.

OLDER AIR CONDITIONERS

Note air conditioners over 10 years old as <u>items requiring repair or replacement within the next 5 years</u> in your inspection report. The compressor will be reaching the end of its lifetime within 5 years.

WORKSHEET

Test yourself on the following questions.
Answers appear on page 130.

1. When is the home inspector required to operate the air conditioning system?

    A. If the outside temperature is under 60° at the time of the inspection and the night before and other conditions are right
    B. If the outside temperature is over 60° at the time of the inspection and the night before and other conditions are right
    C. If the system hasn't been started yet this season and the power hasn't been on for 24 hours
    D. Never

2. Which air conditioning system component removes heat from the pressurized gas and changes the refrigerant to a liquid?

    A. The compressor
    B. The condenser
    C. The expansion device
    D. The evaporator

3. When should the home inspector turn the air conditioner off immediately?

    A. If it starts to rain
    B. If the condenser fan goes on with the compressor
    C. If the compressor groans or squeals
    D. If the condensate tray is rusted

4. What cooling capacity is required for every 550 square feet in a home?

    A. 1 ton
    B. 12,000 tons
    C. 1 BTU
    D. 7 amps

5. What is the function of the evaporator coil?

    A. To compress refrigerant gas and raise its temperature
    B. To absorb heat into the refrigerant liquid and change it to a gas
    C. To release heat from refrigerant gas and change it into a liquid
    D. To reduce pressure on the refrigerant liquid

6. Which refrigerant line is thinner, warmer, and under high pressure?

    A. The gas line between the evaporator and the compressor
    B. The liquid line between the condenser and the evaporator

7. What condition may be indicated if there is ice buildup on the evaporator coils?

    A. Loose mountings on the air handler
    B. Too much air flow through the coil
    C. A missing auxiliary drain line
    D. Not enough air flow through the coil

8. What condition would <u>not</u> indicate that the cooling system is low on refrigerant?

    A. Frost on the larger refrigerant line
    B. A temperature differential lower than 15°
    C. A vibrating condenser fan
    D. Bubbles in the refrigerant line

9. What type of evaporator coil is shown here?

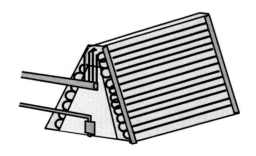

Chapter Twelve

OTHER COOLING SYSTEMS

In this chapter other types of air conditioning systems will be presented.

Water Cooled Air Conditioning

The difference between air cooled air conditioning and water cooled air conditioning is in the **condenser**. All other aspects of the 2 types of systems are the same. A water cooled system condenser uses water from many different sources (a well, lake or pond, recirculated ground-loop) which absorbs heat from the refrigerant. The condenser has a water jacket surrounding the refrigerant coil. The surrounding water cools the refrigerant gas and changes it to a liquid, serving the same purpose as the condenser in an air cooled system.

Because a water cooled system doesn't depend on outside air, the compressor/condenser unit is located inside. The compressor/condenser unit is usually located near the furnace, while the evaporator sits in the plenum. But standalone units may have the condenser and compressor in the air handler cabinet. Because the system doesn't need long refrigerant lines connecting the compressor/condenser with the evaporator, it needs less refrigerant.

A water cooled system (with domestic water) is considered wasteful since warm water from the condenser must be continually discharged and replaced with cooler domestic water. This discharged water is not drinkable and is typically discharged to a floor drain, to the lawn, or to a swimming pool. In some areas, local codes prohibit discharge into the public sewer or private septic system. In fact, they may prohibit the use of water cooled air conditioners entirely. The use of wells and ponds or lakes are a much better source of water. You should check your local codes.

In larger, commercial installations of a water cooled air conditioning system, discharge water may be sent to a **cooling tower** on the roof. The water is cooled in the tower by the use of high volume fans and is recycled to the condenser. This water must be treated to prevent scale buildup in the piping and to keep bacteria and yeasts from growing in it. (Legionnaire's disease is a result of contaminated recycled water.) Although

Guide Note

Pages 129 to 136 present information on other types of cooling systems, including the water cooled system, the gas chiller, and the evaporator cooler. The heat pump, another method of cooling, is presented on page 133.

For Your Information

Check your local codes on whether water cooled systems are permitted and on the acceptable means of discharging warm water from the system.

rare, a home may have a cooling tower, but the home inspector is not required to inspect it other than for leaking.

CAUTION: A water cooled system must have water flowing to the condenser when the system is operating or severe damage can result to the unit. The home inspector should check that the water supply valve to the unit is open before starting the system. Also, heed the warnings about not operating the system if its main power source is turned off. It could have turned off due to a malfunction with the system. Check with the homeowner first.

When inspecting the water cooled system, the home inspector should inspect all equipment (except the condenser) as presented on pages 122 to 127. These are additional conditions the home inspector should look for with a water cooled air conditioning system:

- **Leaks:** Check inlet and outlet piping and the condenser water jacket for leaking. Notice any signs of corrosion or rusting from the effect of leaking.

- **Improper discharge:** Try to determine where warm water from the condenser is being discharged or recycled. If the method used is in violation of your local codes, record that information in the inspection report.

- **Extreme temperature differential:** Feel the inlet and outlet piping at the condenser for temperature difference. The differential should be about 15° to 20°. If the difference is too high, it's an indication that too little water is flowing into the condenser (a smaller amount of water has to absorb more heat).

- **Short cycling compressor:** One of the reasons the compressor in a water cooled system may short cycle is that the water pressure may be too high or too low.

Worksheet Answers (page 128)

1.	B
2.	B
3.	C
4.	A
5.	B
6.	B
7.	D
8.	C
9.	A-coil

The Gas Chiller

Gas chillers operate on the same principles as the refrigerant cycle used in conventional air conditioning systems. The basic differences between these systems are as follows:

- Ammonia or lithium bromide is used as the refrigerant instead of Freon.

- Heat is used to drive the system instead of a compressor,

where natural gas is used to produce the heat required.

The operating principles behind the refrigerant cycle in the gas chiller are the following:

1. A **generator** heats a solution of liquid ammonia changing it to a high-temperature, high-pressure gas.

2. A **condenser** air cools the ammonia gas down to a high-pressure liquid state. A **flow restricter** between the condenser and evaporator reduces the pressure and the temperature of the liquid ammonia.

3. The liquid ammonia in the **evaporator tank** absorbs heat from water piped through the tank. Water from a coil over the furnace which has been warmed by household air enters the tank. Cooled water is sent back to the coil to cool the household air. As the liquid ammonia absorbs heat from the water it changes back into a low-pressure gas.

4. An **absorber** mixes the ammonia gas with a solution of hot liquid ammonia and is air cooled until the gas is absorbed into the solution. A high-pressure pump sends the liquid ammonia back to the generator.

GAS CHILLER

The gas chiller uses ammonia as a refrigerant and uses heat to drive the refrigerant cycle.

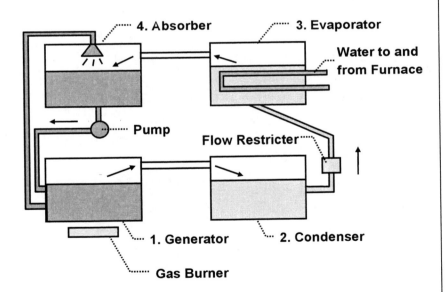

The gas chiller is usually used only in large commercial installations and is not often seen in residential homes. Some smaller home-use chillers have been introduced but have not been very successful. It's rare that the home inspector will come across one.

The Evaporative (Swamp) Cooler

Most air conditioning systems take in warm, moist air and produce cool, dry air. The evaporative cooler works differently, taking in warm, dry air and producing cool, moist air. That's why it's used only in dry climates like the Southwest.

The evaporator cooler consists of a sheet metal or plastic cabinet containing the following:

- A high volume, 2-speed **blower unit** to draw in outside air, pass it through the cooler to cool it, and circulate the cooled air through the duct system. Inspect the blower unit for proper operation — listen for sounds or vibrations indicating loose mountings or blown bearings.

- The **pad type** cooler has **pads** that hold water which evaporates as the warm air passes over it. As the water evaporates it absorbs heat from the air and thus cools it for household use. Pads should be cleaned or replaced regularly to prevent the buildup of salts, algae, bacteria and yeasts. Inspect the pads to see if they need to be cleaned or replaced.

- A **rotary type** cooler has a **drum** made of absorbent screening material that rotates through a tank of water, and the air passes over the top of the drum, causing the water to evaporate.

- **Water supply:** Each type of evaporative cooler has connections to the home's domestic water supply to keep water flowing into the cooler. Coolers have a **pump** to draw water out of a tray to keep the pads wet. Water level in the tray or reservoir is maintained with a float valve that adds water when required.

The main item to watch for when inspecting the evaporative cooler is rusting and leaking of the various parts.

Chapter Thirteen

THE HEAT PUMP

A heat pump is a year-round system that operates as an air conditioner in the hot season and as a heating system in the cold season. It's basically a **reverse-cycle air conditioner**. When it's operating in its heating mode, the evaporator and the condenser switch functions. The evaporator becomes the condenser, producing warm air for the home, and the condenser becomes the evaporator, discharging cool air to the outdoors. This is accomplished by the addition of a **reversing valve** that changes the direction of the refrigerant process.

Air-to-Air Heat Pumps

An air-to-air heat pump is a reverse-cycle air cooled air conditioner. The drawing below shows its operation in the cooling mode. Notice the position of the reversing valve. Here, it sends the returning low-pressure gas in a loop up and then down to the compressor. High-pressure gas leaving the compressor is routed by the valve to the condenser.

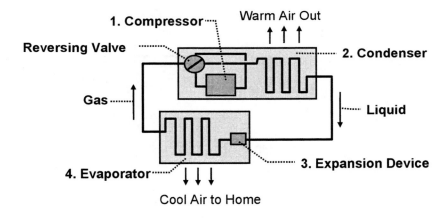

HEAT PUMP IN COOLING MODE

In the winter, the cycle is reversed by changing the position of the reversing valve. The refrigerant actually flows in the opposite direction. Only the function of the coils changes.

1. The high-pressure refrigerant gas leaves the **compressor** and is routed by the reversing valve to the inside coil that now acts as a condenser.

Guide Note

Pages 133 to 136 present the study and inspection of heat pumps for heating and air conditioning.

2. In the **condenser**, located in the furnace plenum or in a standalone attic-mounted unit, heat is removed from the pressurized gas by means of the fan that blows cool house air over it. When the heat is removed, the gas condenses to a warm liquid that is under high pressure.

3. The high-pressure liquid passes through the **expansion device** to reduce its pressure, causing its temperature to drop. (Note that the expansion device is working in a reverse mode too.) This liquid passes to the outside coil that now acts as an evaporator.

4. The **evaporator,** located in the outdoor cabinet, absorbs heat from the liquid refrigerant as outside air is blown over the coils, changing the refrigerant back into a low- pressure gas. Notice the position of the reversing valve —the gas leaving the evaporator is routed by the valve in a loop up and then down to the compressor, where the process begins again.

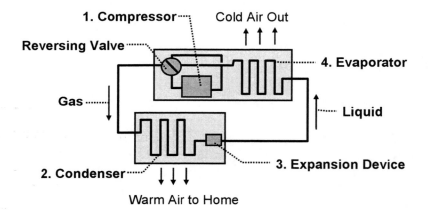

HEAT PUMP IN HEATING MODE

The air-to-air heat pump typically has the same indoor and outdoor components as the air cooled air conditioner. This is the typical **split system** where the compressor, reversing valve, outdoor coil, and fan are in the outdoor cabinet and the indoor unit consists of the inside coil, condensate tray, and condensate line (plus filter and fan if a standalone unit). There is, however, what's called a **triple split system** where the condenser and reversing valve sit in a separate casing located inside the house. The home inspector may find these variations. But the function of all the components remains the same.

The heat pump has its limitations. When the outside temperature falls too low, the outdoor coil loses its ability to collect heat from the outside air and is no longer able to produce enough heat to heat the home. Generally, the heat pump becomes inefficient at temperatures below 35º, depending on the equipment. At a preset temperature the unit automatically shuts down and the furnace (or some other means of auxiliary heating such as electric resistance heaters) must take over the job of heating the home.

Some heat pumps have a gas burner below the outside coil that activates just before the coil would shut down and defrosts the coil. This type of heat pump with its own auxiliary heater will continue operation at low temperatures and does not need additional heat from a furnace or other means.

NOTE: There are also **ground source** and **water source** heat pumps. These systems use pipes buried in the earth below the frost line or located in a body of water such as a lake to collect and dissipate heat. In general, there's more heat available to be collected in winter months than with air-to-air units, but installation costs run much higher. Some local environmental laws prohibit their use. There are also **water-to-air** heat pumps that are reverse-cycle water cooled air conditioners.

Inspecting Air-to-Air Heat Pumps

The home inspector needs to inspect the operation of the heat pump in only one of its modes. **Don't try to run it in both modes during the inspection.** It doesn't matter. The equipment is the same, and any defects to be found will become apparent in either mode. If the heat pump runs in one mode, it will run in the other. In fact, running the heat pump in the wrong mode can result in damage to the compressor.

If the heat pump is currently operating as a cooling system, then inspect it in its cooling mode. If the heat pump is operating as a heating system, then inspect it in its heating mode. The key is the outside temperature at the time of the inspection — if **over 65º**, test it in the cooling mode, and if **under 65º**, test it in its heating mode.

When inspecting the heat pump in either mode, follow the instructions given earlier in this guide for inspecting air cooled

> **CAUTION**
>
> Run the heat pump in only one mode during the inspection — the cooling mode if over 65º and the heating mode if under 65º.

air conditioners, both plenum and standalone units (pages 122 to 127). The inspection is basically the same since the components of the systems are similar and have similar problems. There are only a few points to be made:

- Where **snowfall** is heavy, the base of the outdoor heat pump cabinet should sit 1' off the ground.

- Heat pumps are usually **sized** for the cooling load of the house, not the heating load. If they're sized for heating, they'll have too much cooling power.

- **Thermostats** for heat pumps and supplementary heaters sometimes confuse homeowners. They may constantly turn on the supplementary heaters before the heat pump would normally shut off. Explain that when calling for more heat, the thermostat should be raised in only small increments unless they want more heat immediately.

- Just as in the cooling mode, the **temperature differential** should be about 15° to 22° at the inside coil.

- **Defrost cycle:** Moisture can build up on the outside coil during the heating cycle. In cold weather (37° and below for most units), moisture will condense into frost. A defrost cycle occurs approximately every 90 minutes and lasts approximately 10 minutes.

Reporting Your Findings

When reporting on the inspection of the heat pump, you'll be filling in the sections of your report for both the air conditioning systems and the heating system. Indicate in both sections that the type of system is a heat pump. Refer to earlier pages in this guide for reporting on heating and cooling systems. Be sure to report your findings on the following:

- Brand name, model, and serial number of the unit
- Condition of the heat pump components
- Its operation
- Any conditions that might warrant a major defect or repair.

EXAM

A Practical Guide to Inspecting Heating and Cooling has covered a lot of material. Take the time to test yourself and see how well you've absorbed it.

To receive Continuing Education Units:
Complete the following exam by filling in the answer sheet found at the end of the exam. Return the answer sheet along with a $50.00 check or credit card information to:

American Home Inspectors Training Institute
N19 W24075 Riverwood Dr., Suite 200
Waukesha, WI 53188
Please indicate on the answer sheet which organization you are seeking CEUs.

It will be necessary to pass the exam with at least a 75% passing grade in order to receive CEUs.

Roy Newcomer

Name_____

Address_____

_____

Phone:_____

e-mail:_____

Credit Card #:_____

Exp Date:_____

Fill in the corresponding box on the answer sheet for each of the following questions.

1. Which action is required during the heating inspection according to most standards of practice?

   A. Required to inspect the heat distribution system
   B. Required to inspect the uniformity or adequacy of the heat supply to rooms
   C. Required to operate automatic safety controls
   D. Required to observe the interior of flues and fireplace insert flue connections

2. Which of the following is <u>not</u> a part of the heating inspection?

   A. Heating equipment
   B. Combustion product and disposal system
   C. Humidifiers
   D. Old steam boilers

3. What is the overall purpose of the heating inspection?

   A. To identify the brand name of the heating unit
   B. To verify a heat source in each room
   C. To open all access panels on equipment
   D. To identify major deficiencies in the heating system

4. What might happen if the batteries are low in a programmable thermostat and the inspector turns off the heating unit's safety switch?

   A. The heating unit won't turn off

   B. The thermostat's program can be lost

   C. The thermostat's mercury switch will break

   D. The thermostat will have to be replaced

5. At what temperature should the home inspector reset the thermostat after testing the heating equipment?

   A. To 70°
   B. To the temperature the home inspector thinks is most comfortable for the home
   C. To the temperature the home inspector thinks is most efficient for the heating equipment
   D. To the owner's original setting

6. What is not a function of the automatic gas valve on a gas burner?

   A. To respond to the thermostat to start and stop gas flow to the burners
   B. To sense if the pilot is lighted or not
   C. To respond to the limit control to stop gas flow to the burners during a malfunction
   D. To meter the gas stream for volume and pressure to the burners

7. What situation at the gas burner should cause the inspector to suspect a faulty heat exchanger?

   A. Rusting, flaking, and corrosion in the burner area
   B. A red tag on the gas line to the burner
   C. Flames roaring or dancing on the burners
   D. A faulty thermocouple

8. Yellow-tipped flames at a gas burner are a sign of carbon monoxide production.

   A. True
   B. False

9. Identify the gas burner components marked 1, 2, 3 in the drawing

   A. Gas manifold, ribbon ports, burner tubes

   B. Combustion air shutter, ribbon ports, burner tubes

   C. Gas manifold, burner tubes, ribbon ports

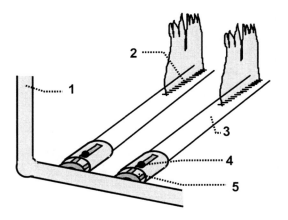

10. What conditions of a gas furnace should be reported as a safety hazard?

   A. Flashback, dirty burners
   B. Spillage from the draft diverter, use of a metal chimney
   C. Blocked flue, dirty burners
   D. Spillage from the draft diverter, blocked flue

11. What color should flames be in a properly operating oil burner?

   A. Blue
   B. White or red
   C. Bright orange or yellowish white
   D. Any color as long as they have sooty or smoky edges

12. Where is an oil burner's primary control located?

   A. In the blast tube or in the oil supply line
   B. On the exhaust stack or on the burner housing

13. What condition would indicate an oil burner

needs an adjustment?

A. A sealed porthole
B. Oily smoke smells or soot
C. Spillage at the barometric damper
D. Cracks or open joints at the firebox

14. Identify the oil components marked, 1, 2, 3 in the drawing

A. Vent pipe, fill pipe, barometric damper

B. Fill pipe, vent pipe, oil burner

C. Vent pipe, fill pipe, oil tank

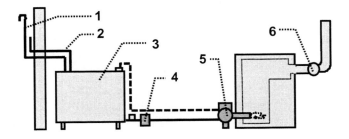

15. What condition is indicated if a lighted match held at the barometric damper leans inward toward the smoke pipe?

A. The damper is broken.
B. A downdraft is occurring.
C. Carbon monoxide is escaping into the home.
D. Exhaust is moving up the chimney as it should be.

16. What type of heating system is shown in Photo #14?

A. A steam boiler
B. A high-efficiency forced warm air furnace
C. A gravity warm air furnace
D. A hydronic boiler

17. Why should the furnace in photo 1 not be operated?

A. Too much aluminum foil present to properly inspect it
B. Improper venting
C. A suspect chimney
D. Rusted ductwork

18. Which item does a conventional forced warm air furnace and a mid-efficiency furnace have in common?

A. A heat exchanger
B. An induced draft fan
C. An intermittent pilot
D. A motorized vent damper

19. The 3 main controls on a furnace are:

A. The thermostat, the gas valve, and the thermocouple.
B. The thermostat, the primary control, and the photocell.
C. The thermostat, the limit control, and the fan control.
D. The thermostat, the limit control, and the pump control.

20. Which photos show signs that are an indication of a cracked heat exchanger?

A. Photos 5,21
B. Photos 11,20
C. Photos 5,25
D. Photos 20,21

21. Why should an open return within 10' of the furnace be reported as a safety hazard?

A. It can cause back drafting at the chimney.
B. It can pull leaking exhaust into the air supply.
C. It can restrict air flow to the furnace.
D. It can damage the heat exchanger.

22. In which furnace can the home inspector expect to be able to inspect the <u>greatest portion</u> of the heat exchanger?

    A. A gas pulse furnace
    B. A high-efficiency gas furnace
    C. A conventional oil furnace
    D. A conventional gas furnace

23. Which of the following should be reported as an item needing repair or replacement within the next 5 years?

    A. Any furnace over 10 years old
    B. Any copper boiler over 10 years old
    C. Any cast iron boiler over 20 years old
    D. Any steel boiler over 25 years old

24. At what pressure would the pressure relief valve open on a forced hot water boiler?

    A. 2 psi to3 psi
    B. 12 psi to 15 psi
    C. 28 psi to 30 psi
    D. Water in a closed system is not under pressure.

25. Upon a call for heat for the thermostat, which components of a newer hydronic boiler are activated simultaneously?

    A. The burner and the circulating pump
    B. The burner and the limit control
    C. The circulating pump and the pressure relief valve
    D. The circulating pump and the low water cut-off

26. What sign appearing just after a forced hot water system is fired indicates a waterlogged expansion tank?

    A. The circulating pump begins leaking.
    B. The electrical connections begin arcing.
    C. Water begins dripping into the burner area.
    D. Pressure builds up quickly and the pressure relief valve discharges.

27. The home inspector should operate radiator control valves.

    A. True
    B. False

28. The home inspector should trip the pressure relief valve during the boiler inspection.

    A. True
    B. False

29. Identify the hydronic boiler components marked 1, 2, 3 in the drawing

    A. Circular pump, pressure relief valve, gauge
    B. Expansion tank, pressure/temp relief valve, gauge
    C. Expansion tank, pressure relief valve, circular pump
    D. Expansion tank, pressure relief valve, gauge

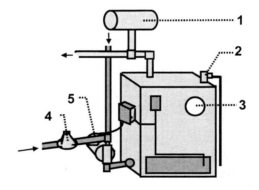

30 How can the home inspector identify a steam boiler?

    A. An expansion tank in the attic
    B. The use of radiators for a heat source
    C. A water level sight gauge on the boiler
    D. A pressure gauge on the boiler

31. What is the normal operating pressure of a steam boiler?

    A. 2 psi to 3 psi
    B. 5 psi
    C. 12 psi to 15 psi
    D. 28 psi to 30 psi

32. What condition should the home inspector suspect if there is a loud banging or knocking just after the steam boiler fires?

   A. The water contains rust and mud.
   B. The Hartford loop is malfunctioning.
   C. The limit control is faulty.
   D. The pipes are not properly sloped.

33. Under what conditions should the home inspector not fire up the steam boiler?

   A. A high water level and a low rust line in the sight gauge
   B. No water in the sight gauge
   C. Dirty water in the sight gauge
   D. Sparkling water in the sight gauge

34. Under what conditions should the home inspector turn off the steam boiler immediately?

   A. Rapidly fluctuating water level
   B. Pressure at 15 psi or more
   C. Evidence of a cracked heat exchanger
   D. Improper relief valve extension

35. What type of air conditioner is the home inspector NOT required to inspect?

   A. Heat pumps
   B. Air cooled central air conditioning
   C. Water cooled central air conditioning
   D. Window air conditioners

36. Which action is required during the cooling inspection according to most standards of practice?

   A. Required to operate cooling systems under all circumstances
   B. Required to operate cooling systems if the outside temperature is under 60°
   C. Required to inspect the cooling system equipment
   D. Required to observe the adequacy of cool-air supply to the various rooms

37. What air conditioning component functions to put refrigerant gas under a high pressure?

   A. The compressor
   B. The condenser
   C. The expansion device
   D. The evaporator

38. A temperature differential of over 25 degrees between the supply and return sides of the evaporator is a sign of something wrong with the cooling system.

   A.   True
   B.   False

39. Should the home inspector turn off a cooling system if the condenser fan never comes on?

   A. Yes
   B. No

40. What installation in the attic mounted cooling system is NOT ALLOWED in most communities?

   A. Having an auxiliary condensate drip pan
   B. Combining the main and the auxiliary condensate lines of two separate units
   C. Draining the condensate lines into the plumbing vent stack
   D. Having a trap on the condensate lines

41. Under what conditions should the home inspector recommend a cooling system be serviced by a technician?

   A. If there's no exterior disconnect switch
   B. If the compressor short cycles
   C. If the condensate line is leaking
   D. If the ductwork is noisy

42. In what location might the evaporator cooler be found?

   A. Florida
   B. Minnesota
   C. Arizona
   D. New Hampshire

43. In which direction would refrigerant in the air-to-air heat pump shown flow when operating in the heating mode

   A. Clockwise

   B. Counter-clockwise

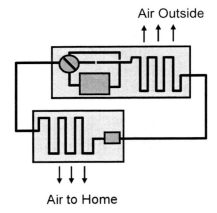

Air Outside

↑ ↑ ↑

↓ ↓ ↓

Air to Home

44. When inspecting a heat pump, the home inspector should:

   A. Test it in both modes

   B. Test it in the heating mode if the temperature is over 65 degrees

   C. Test it in the cooling mode if the temperature is under 60 degrees

   D. Test it in the heating mode if the temperature is under 65 degrees

45. What device on a heat pump allows it to operate in either a cooling or heating mode?

   A. The compressor
   B. The evaporator and condenser coils
   C. The reversing valve
   D. A gas burner under the outside coil

46. The home inspector should turn on a heat pump if the main power supply has been shut off.

   A. True
   B. False

47. **Case study:** You are inspecting a heating system as shown in Photo #9 in this guide. What type of system is it?

   A. A gravity hot water system
   B. A forced hot water system

48. What is the object at the lower right of the photo?

   A. The circulating pump

   B. The expansion tank

   C. The automatic fill, pressure reducing valve

   D. A convector

49. In the case study in number 47, what is the fuel source for the heating unit?

   A. Coal
   B. Gas
   C. Oil

50. In the case study in number 47, which of following conditions found during the inspection should be reported as a safety hazard.

   A. Pressure gauge reading of 30 PSI, leak from the boiler jacket, missing relief valve extension

   B. Leaks from the boiler jacket, hole in the smoke pipe, noisy circulating pump

   C. Pressure gauge reading of 30 PSI, missing relief valve extension, hole in the smoke pipe

   D. Leak from boiler jacket, missing relief valve extension, squeaks in the pipes

GLOSSARY

A-coil A type of evaporator coil in a cooling system that consists of double plates of coils connected at the top, forming an A-shape.

Air cooled air conditioning A cooling system where heat is removed from the refrigerant by means of a fan blowing air over the condenser.

Anticipator A device on a thermostat that turns a heating system on and off just before present temperature settings are reached.

Aquastat A temperature-sensitive device in a boiler that is immersed in water to detect temperature changes in water, thereby activating the circulating pump.

Automatic fill valve A valve on the plumbing line to a boiler that adds water to the system.

Auxiliary condensate line Drain piping from an air conditioning system in addition to the main condensate line.

Balancing duct damper A damper located in the branch supply ducts that equalizes the flow of warm air from the furnace to the house.

Barometric damper A hinged plate located in the smoke pipe above an oil-fired heating system which swings open or closed to regulate drafts.

Bi-metallic vent damper A damper in a smoke pipe that expands open when exhaust gases heat up and contracts closed when gases cool.

Blast tube The component of an oil burner that holds the nozzle and ignition electrodes and extends into the combustion chamber.

Boiler A heating unit that heats the home with hot water or steam distributed through pipes.

British Thermal Unit A unit of measure of heat output, representing the amount of heat required to raise or lower the temperature of 1 pound of water 1° Fahrenheit. Abbreviation *BTU*.

Carbon monoxide A life-threatening gas and combustion byproduct. Abbreviation *CO*.

Circulating pump The pump that circulates water in a hot water heating system.

Closed system A hot water system with a sealed expansion tank located just above the boiler.

Combustion air The air required for mixing with fuel such as oil or gas before the fuel is burned.

Combustion chamber The chamber in a furnace or boiler in which fuel is burned. Called a firebox in older oil burners.

Compressor The component of a cooling system that moves refrigerant and pressurizes the refrigerant gas in order to raise its temperature.

Condensate line Drain piping from an air conditioning system to dispose of condensation.

Condensate tray A tray under an air conditioning unit to collect condensation.

Condenser A coil in a cooling system through which the refrigerant gas flows that removes heat from the gas, condensing it into liquid form.

Condensing units See *High-efficiencies*.

Conduction The transfer of heat from a warmer object to a cooler object by physical contact.

Convection The transfer of heat through air, water, or steam that moves heat from a warmer location to a cooler location.

Convectors Heat outlets in the home consisting of finned plates or pipes that deliver hot water from a boiler and warm the air passing over them.

Conventional heating systems Generally refers to furnaces and boilers manufactured before the mid 1970's which had low operating efficiencies.

Damper A mechanical device used with heating systems to regulate the movement of air through vent piping or supply ducts.

Draft air The air required to insure discharge of exhaust gas from a heating system through the flue.

Draft diverter A device built into a furnace that protects the heating system from excessive updrafts and chimney downdrafts.

Draft hood A device on top of a boiler that protects the heating system from excessive updrafts and chimney downdrafts.

Dry base boiler A boiler with the heat exchanger set above the combustion chamber.

Electric plenum heater An auxiliary heater, usually added to an oil-fired furnace, located in the plenum.

Electric resistance heating Baseboard convectors, wall mounted strips, or floor inserts that each operate as a separate heating plant.

Electronic vent damper A damper in a smoke pipe that automatically opens before the burner starts and closes when exhaust gases cool down.

Equalizer A loop of piping running from the top of a steam boiler to the bottom that balances pressure above and below the water line in the boiler.

Evaporator A coil in a cooling system through which refrigerant flows that causes the refrigerant liquid to absorb heat which changes the liquid to a gas.

Evaporator cooler A cooling system that uses the evaporation process to cool and moisturize the air for home cooling.

Expansion device A device in a cooling system through which refrigerant liquid flows that depressurizes the liquid and cools it down.

Expansion tank A tank in a hot water heating system that provides space for water to expand into. Tanks may be open or closed.

Fan control A temperature-sensitive switch that turns the furnace fan off and on at preset high temperatures.

Fan/limit switch A furnace control combining both a fan control and a limit control.

Fan safety switch A switch on a furnace that prevents the fan from operating when the access panel is removed.

Firebox See *Combustion chamber*.

Flame retention burner An oil burner with a flame retention ring that provides for more efficient combustion.

Flame retention ring On an oil burner, a cone-shaped, slotted head on the blast tube that provides pressure, velocity, and rotation to the air stream.

Flame rollout A condition in a gas burner where the flames burn outside the combustion chamber.

Forced hot water heating system A system in which water heated in a boiler is circulated by means of a circulating pump through pipes to radiators or convectors throughout the house. Also called a hydronic system.

Forced warm air heating system A system in which air heated in a furnace is circulated by means of a blower unit through ductwork to registers in the house.

Furnace A heating unit that heats the home with warm air distributed through ductwork.

Gas chiller A cooling system that uses ammonia as a refrigerant and uses heat to drive the refrigerant cycle.

Gas manifold In a gas burner, the gas line that delivers gas to the burners.

Gravity hot water heating system A system in which water heated in a boiler rises naturally through pipes to radiators in the house.

Gravity steam heating system A system in which water boiled in a boiler changes to steam which rises naturally through pipes to radiators and condensate drains back to the boiler.

Gravity warm air heating system A system in which air heated in the furnace rises naturally through ductwork without the aid of a blower unit. Also called an octopus furnace.

Hartford loop A loop of return piping that turns downward to connect to the equalizer just below the water level, preventing water from flowing out of a steam boiler in case of a leak in return piping.

Heat exchanger A heavy metal hood above the combustion chamber that holds and contains the burner flame. In a furnace, it separates exhaust air from the circulating air that heats the house. In a boiler, it separates exhaust air from the circulating water.

Heat pump A reverse-cycle air conditioner that operates as both a cooling and heating system by reversing the flow of refrigerant.

High-efficiencies Gas and oil furnaces and boilers with an operating efficiency in the 95% range. Units typically have two or more heat exchangers, reducing exhaust gas temperatures low enough to condense. Also called condensing units.

Humidifier A device added to a furnace that adds moisture to the circulating air.

Hydronic boiler A boiler used in a forced hot water heating system.

Ignition electrodes On an oil burner, a pair of electrical elements located in the blast tube to provide a spark that ignites the oil.

Induced draft fan A fan that pulls combustion byproducts through a furnace or boiler, ensuring a good draft and reducing heat loss.

Intermittent pilot A pilot which lights only on a call for heat.

Limit control A control that turns off the burners in a furnace when circulating air reaches a preset high temperature or in a boiler when water reaches a preset high temperature (forced hot water) or pressure reaches a preset high pressure (steam).

LP gas Liquefied propane or petroleum gas used as a fuel source. Can also be a mixture of propane, butane, or iso-butane.

Low water cut-off A safety control on a boiler that shuts off the burner when water levels fall below the required amount for the system.

Main automatic gas valve A device on a furnace or boiler that starts and stops the flow of gas to the burners.

Manual turn-off A valve that turns off the fuel supply to a heating unit.

Master shut-off A device that turns off electricity to the heating system and its controls. Also called a safety switch or serviceman's switch.

Mid-efficiencies Gas and oil furnaces and boilers with an operating efficiency in the 80% range. Units typically have an induced draft fan and motorized vent dampers.

Modulating aquastat A device in a hot water boiler that senses outdoor temperature and changes circulating water temperature requirements accordingly.

Motorized duct damper A damper located in zoned supply ducts that controls the flow of warm from a furnace to zones within the house.

Motorized vent damper A damper located in the smoke pipe above a furnace or boiler that opens and closes automatically to prevent heat loss up the chimney.

Octopus furnace See *Gravity warm air heating system*.

Oil burner nozzle A device located in the blast tube which shoots out oil particles into the firebox.

Open system A hot water system that has an expansion tank open to the atmosphere and located above the highest radiator in the home.

1-pipe system A hot water or steam heating system with a single piping run to and from the boiler and the radiators or convectors.

Open return In the basement, an opening in the furnace or return ducts for collecting return air. Most often prohibited.

Overshooting A condition when a house is heated higher than the thermostat is set for.

Pilot The small flame that ignites gas in a gas burner.

Plenum The first large section of supply duct directly over a furnace from which smaller ducts branch out to distribute heat to the house.

Pressure reducing valve A valve on the water supply line to a boiler that reduces water pressure to an acceptable boiler pressure.

Pressure relief valve A valve on a boiler that will discharge water from the boiler if pressure approaches dangerous limits.

Primary control A control on a oil burner that starts and stops the burner in response to thermostat signals and verifies ignition. May be located on the smoke pipe or burner housing.

Puffback In an oil burner, a condition where smoke, soot, or flames escape from the combustion chamber.

Pulse unit A furnace or boiler with a combustion method involving a sealed combustion chamber and a tailpipe that create a self-perpetuating series of shockwave ignitions.

Pump control A control on a boiler that turns on the circulating pump at a signal from the thermostat upon a call for heat or from an aquastat at a certain preset temperature.

Radiant heating A system consisting of continuous piping from the boiler or electric cables laid out in rows buried in the floor or ceiling, thereby heating a room by radiation.

Radiation The transfer of heat from a warmer object to a cooler object not in contact with it by radiating heat energy.

Radiators Heat outlets made of heavy metal piping that deliver hot water or steam from the boiler and radiate heat into the room.

Refrigerant The gas in a cooling system that changes between a gas and liquid state, commonly called Freon.

Refrigerant lines The copper piping that moves the refrigerant in both gas and liquid states through a cooling system.

Registers Heat outlets in the home that deliver warm air from the furnace.

Return ducts Ducts that deliver cool air from the home back to the furnace.

Reversing valve A valve in a heat pump that changes the direction of the refrigerant process.

Safety switch See *Master shut-off*.

Serviceman's switch See *Master shut-off*.

Setback feature A feature on a thermostat that can be set to automatically lower temperature settings during certain hours.

Short cycling A condition where the heating or cooling system turns on and off too often.

Spillage A condition where combustion byproducts spill out of a heating unit's exhaust system.

Stack relay A oil burner primary control that is located on the smoke pipe.

Standalone air conditioning system A cooling system with a compressor/condenser unit and an evaporator unit with its own fan, filter, and ductwork. Does not work in conjunction with a furnace.

Standing pilot A pilot flame that burns continuously.

Supply ducts Ducts that deliver warm air from a furnace to the home or deliver cool air from a cooling system to the home.

Tankless coil A coil inserted into a boiler to heat water for domestic use.

Temperature, pressure gauge A gauge on a boiler that shows the current operating temperature and pressure in the boiler.

Thermocouple A bimetallic element that senses whether or not the pilot is lighted and controls the pilot control valve, turning off the flow of gas to the pilot when the pilot flame is out.

Thermostat A temperature-sensitive device that opens and closes a circuit in response to temperature changes in the air, thereby activating a heating or cooling system.

2-pipe system A hot water or steam heating system with separate supply and return piping from the boiler to the radiators or convectors.

Water cooled air conditioning A cooling system where heat is removed from the refrigerant by means of water surrounding the condenser.

Water level sight gauge A glass tube showing water level in a steam boiler.

Wet base boiler A boiler where the heat exchanger surrounds the combustion chamber.

Zone control Where different areas of the home are under the control of different thermostats

INDEX

A Practical Guide to Inspecting Program
Study Unit Six, Inspecting Heating and Cooling

Student Name: _____ Date: _____

Address: _____

Phone: _____ Email: _____

Organization obtaining CEUs for: _____ Credit Card Info: _____

After you have completed the exam, mail *this exam answer page* to American Home Inspectors Training Institute. You may also fax in your answer sheet. You will be notified of your exam results.

Fill in the box(es) for the correct answer for each of the following questions:

1. A☐ B☐ C☐ D☐	24. A☐ B☐ C☐ D☐	47. A☐ B☐
2. A☐ B☐ C☐ D☐	25. A☐ B☐ C☐ D☐	48. A☐ B☐ C☐ D☐
3. A☐ B☐ C☐ D☐	26. A☐ B☐ C☐ D☐	49. A☐ B☐ C☐
4. A☐ B☐ C☐ D☐	27. A☐ B☐	50. A☐ B☐ C☐ D☐
5. A☐ B☐ C☐ D☐	28. A☐ B☐	
6. A☐ B☐ C☐ D☐	29. A☐ B☐ C☐ D☐	
7. A☐ B☐ C☐ D☐	30. A☐ B☐ C☐ D☐	
8. A☐ B☐	31. A☐ B☐ C☐ D☐	
9. A☐ B☐ C☐	32. A☐ B☐ C☐ D☐	
10. A☐ B☐ C☐ D☐	33. A☐ B☐ C☐ D☐	
11. A☐ B☐ C☐ D☐	34. A☐ B☐ C☐ D☐	
12. A☐ B☐	35. A☐ B☐ C☐ D☐	
13. A☐ B☐ C☐ D☐	36. A☐ B☐ C☐ D☐	
14. A☐ B☐ C☐	37. A☐ B☐ C☐ D☐	
15. A☐ B☐ C☐ D☐	38. A☐ B☐	
16. A☐ B☐ C☐ D☐	39. A☐ B☐	
17. A☐ B☐ C☐ D☐	40. A☐ B☐ C☐ D☐	
18. A☐ B☐ C☐ D☐	41. A☐ B☐ C☐ D☐	
19. A☐ B☐ C☐ D☐	42. A☐ B☐ C☐ D☐	
20. A☐ B☐ C☐ D☐	43. A☐ B☐	
21. A☐ B☐ C☐ D☐	44. A☐ B☐ C☐ D☐	
22. A☐ B☐ C☐ D☐	45. A☐ B☐ C☐ D☐ E☐	
23. A☐ B☐ C☐ D☐	46. A☐ B☐	